Distributed and economic model predictive control: beyond setpoint stabilization

Von der Fakultät Konstruktions-, Produktions- und Fahrzeugtechnik
und dem Stuttgart Research Centre for Simulation Technology
der Universität Stuttgart zur Erlangung der Würde eines
Doktor-Ingenieurs (Dr.-Ing.) genehmigte Abhandlung

Vorgelegt von

Matthias A. Müller

aus Stuttgart

Hauptberichter: Prof. Dr.-Ing. Frank Allgöwer
Mitberichter: Prof. Dr. Lars Grüne
Prof. Dr. David Angeli

Tag der mündlichen Prüfung: 25. Juli 2014

Institut für Systemtheorie und Regelungstechnik

Universität Stuttgart

2014

Bibliografische Information der Deutschen Nationalbibliothek

Die Deutsche Nationalbibliothek verzeichnet diese Publikation in der
Deutschen Nationalbibliografie; detaillierte bibliografische Daten sind
im Internet über http://dnb.d-nb.de abrufbar.

D93

ISBN 978-3-8325-3821-7

Logos Verlag Berlin GmbH
Comeniushof, Gubener Str. 47,
10243 Berlin
Tel.: +49 (0)30 42 85 10 90
Fax: +49 (0)30 42 85 10 92
INTERNET: http://www.logos-verlag.de

Für Judith

Acknowledgements

The are many people who supported me during my time as a PhD student and to whom I would like to express my gratitude. First of all, I thank my advisor Prof. Frank Allgöwer for providing me with lots of freedom to pursue different research ideas, for all the motivating and encouraging comments and the valuable advice. I am grateful for the fruitful research environment he provided, with many international contacts and the opportunity to attend various conferences in order to share and discuss my work with other researchers in the field.

I also want to thank Prof. David Angeli (Imperial College London and University of Florence) and Prof. Lars Grüne (University of Bayreuth) for their interest in my work, for many fruitful discussions and the good collaboration over the past years, and for being members of my doctoral examination committee. In particular, I am indebted to David for hosting me for three months at the Imperial College in 2012; this research stay had a great influence and served as a starting point for various of the results presented in this thesis.

Furthermore, I am grateful to Prof. Daniel Liberzon (University of Illinois at Urbana-Champaign). Although he was not directly involved in the research which led to this thesis, I benefited a lot from the advice he gave me when supervising my master thesis, and I also very much enjoyed our continued collaboration during the last years on the subject of norm-controllability.

I am very thankful to all of my colleagues at the Institute for Systems Theory and Automatic Control (IST) at the University of Stuttgart. You made my time here at the IST very enjoyable and I look forward to our future interaction. In particular, I would like to thank my long-time office mate Mathias Bürger and all the former and current members of the MPC group for many interesting discussions and the good collaboration: Florian Bayer, Christoph Böhm, Florian Brunner, Christian Feller, Gregor Goebel, Christoph Maier, Martin Löhning, Matthias Lorenzen, Marcus Reble, and Shuyou Yu.

I also want to thank my parents for all their love and support during my whole life, and my five siblings and many friends for all the encouragement and the good times we have had together. Heartfelt thanks go to my wonderful wife Judith. Obviously, I had the most important meeting during my whole PhD time already on my third day - which was when I first met her. Thanks for being such a loving, supporting and inspiring partner for me.

Finally, as I love making music, I would like to close this acknowledgement section by quoting one of the most famous composers of all times, Johann Sebastian Bach. He signed many of his compositions with "S.D.G.", which stands for the Latin phrase *Soli Deo Gloria*, and with which he expressed his gratitude to the one who enabled him to do his work (see, e.g., Rueger (1985)).

Stuttgart, July 2014
Matthias A. Müller

Table of Contents

Notation

In the following, we specify the notation used throughout this thesis. Some additional notation required in specific sections is defined at the respective places.

$\mathbb{R}$	set of real numbers
$\mathbb{R}_{\geq 0}$	set of nonnegative real numbers
$\mathbb{R}_{\leq 0}$	set of nonpositive real numbers
$[a, b]$	interval $\{x \in \mathbb{R} : a \leq x \leq b\}$ for constants $a, b \in \mathbb{R}$
$\mathbb{I}$	set of integer numbers
$\mathbb{I}_{\geq a}$	set of integer numbers greater than or equal to $a \in \mathbb{R}$
$\mathbb{I}_{[a,b]}$	set of integer numbers in the interval $[a, b] \subseteq \mathbb{R}$
$\lceil a \rceil$	smallest integer greater than or equal to $a \in \mathbb{R}$
$\lfloor a \rfloor$	largest integer smaller than or equal to $a \in \mathbb{R}$
I_n	$n \times n$ identity matrix
$0_{n \times m}$	$n \times m$ matrix of zeros
A^T	transpose of a matrix $A \in \mathbb{R}^{n \times m}$
$\lambda_{\min}(A)$	minimum real part of the eigenvalues of a matrix $A \in \mathbb{R}^{n \times n}$
$\lambda_{\max}(A)$	maximum real part of the eigenvalues of a matrix $A \in \mathbb{R}^{n \times n}$
$\mathrm{diag}(A_1, \ldots, A_p)$	block-diagonal matrix with main diagonal blocks $A_1, \ldots, A_p$ for some $p \in \mathbb{I}_{\geq 0}$
$\otimes$	Kronecker product of two matrices, see (Horn and Johnson, 1991, p.243)
$A \succ 0 \ (A \succeq 0)$	matrix $A \in \mathbb{R}^{n \times n}$ is positive definite (positive semidefinite), i.e., A is symmetric and $x^T A x > 0$ ($x^T A x \geq 0$) for all $x \in \mathbb{R}^n$ with $x \neq 0$
$A \prec 0 \ (A \preceq 0)$	matrix $A \in \mathbb{R}^{n \times n}$ is negative definite (negative semidefinite), i.e., A is symmetric and $x^T A x < 0$ ($x^T A x \leq 0$) for all $x \in \mathbb{R}^n$ with $x \neq 0$
$:=$	definition, i.e., $a := b$ means that a is defined to be equal to b
$>, \geq, \leq, <$	For vectors $a \in \mathbb{R}^n$, inequalities are intended componentwise, i.e., $a > 0$ means that $a_i > 0$ for all $i \in \mathbb{I}_{[1,n]}$, where $a = [a_1 \ \ldots \ a_n]^T$.
$\lvert x \rvert$	Euclidean norm of $x \in \mathbb{R}^n$
$\lvert x \rvert_{\mathcal{A}}$	Euclidean distance of $x \in \mathbb{R}^n$ to a set $\mathcal{A} \subseteq \mathbb{R}^n$, i.e., $\lvert x \rvert_{\mathcal{A}} := \inf_{a \in \mathcal{A}} \lvert x - a \rvert$
$\mathrm{int}(\mathcal{A})$	interior of a set $\mathcal{A} \subseteq \mathbb{R}^n$
$\overline{\mathcal{A}}$	closure of a set $\mathcal{A} \subseteq \mathbb{R}^n$
$\oplus$	Minkowski sum of two sets $\mathcal{A}, \mathcal{B} \subseteq \mathbb{R}^n$, i.e., $\mathcal{A} \oplus \mathcal{B} := \{a + b : a \in \mathcal{A}, b \in \mathcal{B}\}$

$B_\varepsilon(x)$	ball of radius ε centered at $x \in \mathbb{R}^n$, i.e., $B_\varepsilon(x) := \{z \in \mathbb{R}^n : |z - x| \leq \varepsilon\}$
$B_\varepsilon(x, y)$	$B_\varepsilon(x, y) := B_\varepsilon([x^T \ y^T]^T)$, for $x \in \mathbb{R}^{n_x}$ and $y \in \mathbb{R}^{n_y}$
$\mathcal{K}$	A function $\alpha : \mathbb{R}_{\geq 0} \to \mathbb{R}_{\geq 0}$ is a class $\mathcal{K}$ function (or $\alpha \in \mathcal{K}$ for short), if it is continuous, strictly increasing, and $\alpha(0) = 0$.
$\mathcal{K}_\infty$	A function $\alpha : \mathbb{R}_{\geq 0} \to \mathbb{R}_{\geq 0}$ is a class $\mathcal{K}_\infty$ function (or $\alpha \in \mathcal{K}_\infty$ for short), if $\alpha \in \mathcal{K}$ and $\alpha(r) \to \infty$ for $r \to \infty$.
$\nabla_x F(\bar{x})$	For a function $F(x)$, $F : \mathbb{R}^n \to \mathbb{R}^m$, denote by $\nabla_x F(\bar{x}) \in \mathbb{R}^{m \times n}$ the Jacobian matrix of F, evaluated at the point $\bar{x}$.
$\nabla_x F(\bar{x}, \bar{y})$	For a function $F(x, y)$, $F : \mathbb{R}^{n_1 \times n_2} \to \mathbb{R}^m$, denote by $\nabla_x F(\bar{x}, \bar{y}) \in \mathbb{R}^{m \times n_1}$ the matrix of partial derivatives of F with respect to x, evaluated at the point $(\bar{x}, \bar{y})$.
$\nabla_x^2 F(\bar{x})$	For a function $F(x)$, $F : \mathbb{R}^n \to \mathbb{R}$, denote by $\nabla_x^2 F(\bar{x}) \in \mathbb{R}^{n \times n}$ the Hessian matrix of F, evaluated at the point $\bar{x}$.
$\boldsymbol{v}$	Boldface symbols denote finite sequences $\boldsymbol{v} : \mathbb{I}_{[0,N]} \to \mathbb{R}^n$ for some $N \in \mathbb{I}_{\geq 0}$, or infinite sequences $\boldsymbol{v} : \mathbb{I}_{\geq 0} \to \mathbb{R}^n$, i.e., $\boldsymbol{v} = \{v(0), \ldots, v(N)\}$ or $\boldsymbol{v} = \{v(0), v(1), \ldots\}$, respectively.
$Av[\boldsymbol{v}]$	set of asymptotic averages of a bounded sequence $\boldsymbol{v} : \mathbb{I}_{\geq 0} \to \mathbb{R}^n$, defined by

$$Av[\boldsymbol{v}] := \{\bar{v} \in \mathbb{R}^n : \exists \{t_n\} \subseteq \mathbb{I}_{\geq 0} \text{ s.t. } \lim_{n \to \infty} \frac{\sum_{k=0}^{t_n} v(k)}{t_n + 1} = \bar{v}\},$$

where $\{t_n\}$ is an (arbitrary) infinite subsequence of $\mathbb{I}_{\geq 0}$.

Abstract

In this thesis, we study model predictive control (MPC) schemes for control tasks which go beyond the classical objective of setpoint stabilization. In particular, we consider two classes of such control problems, namely *distributed MPC* for cooperative control in networks of multiple interconnected systems, and *economic MPC*, where the main focus is on the optimization of some general performance criterion which is possibly related to the economics of a system. The contributions of this thesis are to analyze various systems theoretic properties occurring in these type of control problems, and to develop distributed and economic MPC schemes with certain desired (closed-loop) guarantees.

To be more precise, in the field of distributed MPC (see Chapters 3 and 5) we propose different algorithms which are suitable for general cooperative control tasks in networks of interacting systems. We show that the developed distributed MPC frameworks are such that the desired cooperative goal is achieved, while coupling constraints between the systems are satisfied. Furthermore, we discuss implementation and scalability issues for the derived algorithms, as well as the necessary communication requirements between the systems. The proposed distributed MPC schemes cover both the scenarios where (i) the local, individual objective of each system is consistent with the overall goal to be achieved (see Chapter 3) as well as (ii) situations where the systems are self-interested (or competing) and their local, individual performance criteria are possibly conflicting with the overall cooperative goal (see Chapter 5). Particularly important examples for a cooperative control task are consensus and synchronization problems, which have received significant attention in recent years and which appear in many different applications. Here, the systems have to agree on an a priori unknown state (or output) value or trajectory. We exemplarily show how the developed distributed MPC schemes can be used for this special cooperative control problem, which we also illustrate with numerical examples.

In the field of economic MPC (see Chapter 4), the contributions of this thesis are three-fold. Firstly, we thoroughly examine a certain dissipativity condition, which has turned out to play a crucial role in economic MPC. In particular, we analyze its necessity for optimal steady-state operation of a system and its robustness with respect to parameter changes. Secondly, we develop economic MPC schemes which also take average constraints into account. We show how suitable economic MPC schemes using a setting with terminal region and terminal cost can be formulated such that both asymptotic and transient average constraint satisfaction is guaranteed. Furthermore, we provide a Lyapunov-like convergence analysis for the closed-loop system under a relaxed dissipativity condition. Thirdly, we propose an economic MPC framework with self-tuning terminal cost and a generalized terminal constraint, and we show how self-tuning update rules for the terminal weight can be derived such that desirable closed-loop average performance bounds can be established.

Deutsche Kurzfassung

Die prädiktive Regelung (Englisch: *model predictive control*), meist mit MPC abgekürzt, ist ein modernes, optimierungsbasiertes Regelungskonzept mit weitreichender Verbreitung sowohl in der akademischen Forschung als auch in der industriellen Anwendung. Die vorliegende Arbeit befasst sich mit dem Einsatz prädiktiver Regelung in Aufgabenstellungen, die über das klassische Regelziel der Stabilisierung eines vorgegebenen Sollwertes hinausgehen. Insbesondere werden die folgenden zwei Klassen derartiger Problemstellungen betrachtet: verteilte kooperative Regelungsaufgaben in Netzwerken von mehreren, miteinander interagierenden Systemen sowie Regelungsaufgaben, bei denen das Hauptaugenmerk auf der Optimierung eines allgemeinen Gütekriteriums liegt, welches z.B. wirtschaftlichen Gesichtspunkten wie der Profitmaximierung oder Kostenminimierung entspricht. Der Hauptbeitrag der vorliegenden Dissertation besteht in der Analyse verschiedener, für diese Art von Problemstellungen relevanter systemtheoretischer Eigenschaften sowie in der Entwicklung von verteilten und ökonomischen prädiktiven Regelungsverfahren, für die gewünschte Garantien gegeben werden können. Insbesondere werden die folgenden Ergebnisse erzielt.

Im Bereich der verteilten prädiktiven Regelung (s. Kapitel 3 und 5) werden verschiedene Verfahren entwickelt, die für allgemeine kooperative Regelungsaufgaben in Netzwerken von miteinander interagierenden Systemen geeignet sind. Wir zeigen, dass die vorgeschlagenen verteilten MPC-Verfahren die Erfüllung des kooperativen Regelziels sowie die Einhaltung von Zustands- und Eingangsbeschränkungen, die möglicherweise mehrere Systeme miteinander verkoppeln, garantieren. Des Weiteren werden Implementierungsaspekte, die Skalierbarkeit und die notwendigen Anforderungen an die Kommunikationsstruktur zwischen den Systemen diskutiert. Die ersten beiden der entwickelten verteilten MPC-Verfahren sind für Szenarien geeignet, in denen sich die lokalen, individuellen Regelziele der einzelnen Systeme in Einklang mit der gesamten, kooperativen Regelungsaufgabe befinden (s. Kapitel 3), während das dritte in Netzwerken egoistischer, konkurrierender Systeme Anwendung findet, wo die lokalen, individuellen Regelziele möglicherweise in Konflikt mit der zu erreichenden gesamten, kooperativen Aufgabe stehen (s. Kapitel 5). Besonders wichtige spezielle Beispiele einer kooperativen Regelungsaufgabe, die in den letzten Jahren sehr viel Beachtung in der Forschung gefunden haben und auf die viele verschiedene Anwendungen zurückgeführt werden können, sind Konsens- und Synchronisationsprobleme, bei denen sich die Systme auf eine a priori unbekannte Zustands- oder Ausgangstrajektorie einigen müssen. Es wird exemplarisch gezeigt, wie die entwickelten verteilten MPC-Verfahren für diese spezielle kooperative Regelungaufgabe angewandt werden können; dies wird auch durch verschiedene numerische Beispiele illustriert.

Im Bereich der ökonomischen prädiktiven Regelung (s. Kapitel 4) besteht der Beitrag der vorliegenden Arbeit im Wesentlichen aus den folgenden drei Aspekten: Zunächst wird eine Dissipativitätsbedingung, die eine zentrale Rolle für die ökonomische prädiktive Regelung spielt, analysiert. Insbesondere wird ihre Notwendigkeit für den optimalen Betrieb eines Systems an einem Gleichgewichtspunkt sowie ihre Robustheit gegenüber Parameterveränderungen untersucht. Zweitens werden verschiedene ökonomische MPC-Methoden entwickelt, bei denen zusätzlich Beschränkungen auf bestimmte Durchschnittswerte von Eingangs- und Zustandsgrößen berücksichtigt werden. Es wird gezeigt, wie ökonomische MPC-Verfahren mit zusätzlichen Endkosten und -beschränkungen formuliert werden können, die sowohl die Einhaltung von asymptotischen als auch von transienten Durschnittsbeschränkungen garantieren. Des Weiteren wird eine Stabilitätsanalyse für ökonomische prädiktive Regelung mit Durschnittsbeschränkungen durchgeführt, bei der gezeigt wird, dass die Lösung des geschlossenen Regelkreises konvergiert, falls eine relaxierte Dissipativitätsbedingung erfüllt ist. Schließlich wird ein ökonomisches MPC-Verfahren mit selbsteinstellenden Endkosten und einer verallgemeinerten Endbeschränkung entwickelt. Insbesondere wird gezeigt, wie das selbsteinstellende Endgewicht definiert werden kann, um gewünschte Grenzen für die erreichbare durschnittliche Regelgüte des geschlossenen Kreises garantieren zu können.

Chapter 1

Introduction

1.1 Motivation

Model predictive control (MPC) is a very successful modern control technology. It consists of repeatedly solving a finite horizon optimal control problem and then applying the first part of the solution to the considered system. The main advantages of MPC and the reasons for its widespread success are that (i) satisfaction of hard input and state constraints for the closed-loop system can be guaranteed, (ii) optimization of some performance criterion is directly incorporated in the controller design, and (iii) it can be applied to nonlinear systems with possibly multiple inputs (see, e.g., Rawlings and Mayne (2009)). Both the examination of theoretical properties of MPC as well as reports of successful implementations in many different application contexts have received significant attention in the literature. In most of these results, the considered control objective is the stabilization of some given setpoint, and various questions of interest in this context such as closed-loop stability and optimality, or robustness against model uncertainties and disturbances have been studied (see, e.g., Darby and Nikolaou (2012); Findeisen et al. (2003); Mayne et al. (2000)).

In many control problems, however, the desired goal is different from the classical control objective of setpoint stabilization. This is especially true when considering *distributed cooperative* control tasks and control objectives related to the *economics* of a system. These two aspects also define the two main fields considered in this thesis. Concerning the former, a typical setting is, for example, that multiple dynamical systems have to achieve some common cooperative goal, such as multi-vehicle platooning and formation stabilization, or distributed power generation. Cooperative control problems of particular interest due to their relevance in many applications are consensus and synchronization problems, in which multiple systems have to agree on an a priori unknown state or output trajectory. Due to the manifold advantages of MPC mentioned above, a very interesting question is how MPC can be used for the solution of distributed cooperative control tasks. However, as we will argue in the following, there exist only very few distributed MPC schemes which consider such general cooperative control problems, i.e., go beyond the goal of setpoint stabilization. One of the main contributions of this thesis is to develop new solutions for this to a great extent open problem and to provide several distributed MPC schemes which are suited for various cooperative control problems.

The second field considered in this thesis is economic MPC. Here, some general objective is considered which is possibly related to the economics of the system, such as maximization of a certain product in the process industry, minimization of energy consumption, or maximization of the achievable profit in general. When using such a general objective func-

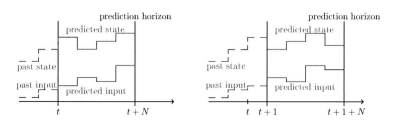

Figure 1.1: Basic idea of model predictive control.

tion within an MPC framework, the resulting closed-loop system may not be converging to some steady-state, but other non-constant trajectories might exist yielding superior performance. The second main contribution of this thesis is to study several systems theoretic questions and properties arising in these type of control problems and to develop novel economic MPC schemes with desired closed-loop guarantees. This includes a classification under what conditions it is optimal to operate a system at some steady-state and when this is not the case, the consideration of *average* constraints besides standard pointwise-in-time constraints, and the development of an economic MPC scheme with a self-tuning terminal cost.

In summary, the goal of this thesis is to develop a deepened systems theoretic understanding of economic and distributed cooperative control problems with an objective beyond setpoint stabilization, as well as the development of several MPC schemes for their solution. In the following two sections, we provide an overview on related work in this research area, and we further motivate and specify in more detail the contributions obtained in this thesis.

1.2 Research topic overview

In this section, we provide a brief overview on the field of model predictive control, with a particular focus on distributed and economic MPC. For a more detailed exposition and several other aspects of MPC including various application examples, we refer the interested reader to (Camacho and Bordons, 2004; Darby and Nikolaou, 2012; Findeisen et al., 2003; Grüne and Pannek, 2011; Mayne et al., 2000; Qin and Badgwell, 2003; Rawlings and Mayne, 2009). The basic idea of model predictive control is sketched in Figure 1.1. Namely, at each sampling instant t, the future behavior of the system is predicted over some finite horizon N using some prediction model, and an open-loop optimal control problem is solved to determine the optimal input trajectory over this time horizon. Then, the first part of this optimal input is applied to the system until the next sampling instant $t + 1$, at which the horizon is shifted and the whole procedure is repeated again.

In classical MPC schemes, the cost function used within the optimal control problem is typically assumed to be positive definite with respect to some setpoint to be stabilized. However, in order to guarantee closed-loop stability, additional assumptions are needed in general. The probably most widely used approach in the literature is to impose additional terminal constraints in the optimization problem and/or use an additional terminal cost term in the cost functional. As shown in more detail in Section 2.1, asymptotic stability can be guaranteed if the terminal constraint and the terminal cost are suitably defined (Chen and Allgöwer, 1998; Findeisen et al., 2003; Fontes, 2001; Grüne and Pannek, 2011; Mayne et al., 2000; Rawlings and Mayne, 2009). Alternative ways to ensure stability are Lyapunov-based MPC methods (Mhaskar et al., 2006), where a global control Lyapunov function is assumed to be known, and so-called *unconstrained* MPC schemes without terminal constraints (Grüne, 2009; Grüne and Pannek, 2011; Reble, 2013; Reble and Allgöwer, 2012), which typically require additional controllability conditions. In recent years, various properties of stabilizing MPC schemes have been studied, such as inherent robustness (Pannocchia et al., 2011; Yu, 2011) and robust MPC approaches (Limon et al., 2009; Mayne et al., 2005; Raimondo et al., 2009; Raković et al., 2012; Yu, 2011), or implementation under unreliable communication in networked control systems (Findeisen et al., 2011; Grüne et al., 2014; Pin and Parisini, 2011). Moreover, stabilizing MPC schemes have been developed for different special system classes such as switched and hybrid systems (Lazar et al., 2006; Mhaskar et al., 2005; Müller et al., 2012a), periodic systems (Böhm, 2011; Gondhalekar and Jones, 2011) or time-delay systems (Reble, 2013). Furthermore, control objectives different from setpoint stabilization such as tracking and path following problems have been studied (Faulwasser, 2013; Faulwasser and Findeisen, 2009; Ferramosca et al., 2009; Limon et al., 2008). In the following, we will more closely review available results in the fields of *distributed* and *economic* MPC, which constitute related work to the results developed in this thesis.

1.2.1 Distributed MPC

The development of distributed MPC schemes has received significant attention in recent years and a large number of publications is available concerning different settings and applications. The following brief overview is by no means exhaustive and mainly considers those works which are important for and closest related to the results presented in this thesis. For a more comprehensive overview of available distributed MPC schemes and many more references, the interested reader is referred to the survey articles (Christofides et al., 2013; Scattolini, 2009) and the recent edited volume (Maestre and Negenborn, 2014).

One possible classification of distributed MPC schemes, which was, e.g., also adopted by Maestre and Negenborn (2014), is to consider what underlying perspective is taken on the overall system to be controlled. Namely, a first perspective is to start with one large-scale system, consisting of many (physically) interconnected components, which is, e.g., the case in large chemical processes, (water) distribution networks, etc. In such settings, computing a centralized controller for the overall system considering some overall (centralized) objective is in general not tractable, as not enough information and/or computational power is available in a single node. Instead, the overall control problem is decomposed and distributed controllers are determined in each subsystem. For such a setting, different distributed MPC schemes have been proposed, for example iterative methods based on distributed optimization and negotiation techniques (Conte et al., 2012; Doan et al., 2011; Giselsson and Rantzer, 2014; Groß and Stursberg, 2013; Kögel and Findeisen, 2012; Maestre et al., 2011; Scheu and Marquardt, 2011; Stewart et al., 2010; Venkat et al., 2005), or schemes based on robustness concepts (Farina and Scattolini, 2012), dissipativity (Varutti et al., 2012) and Lyapunov-based methods (Heidarinejad, 2012; Liu et al., 2010).

On the other hand, a different perspective is to consider a (possibly large) number of dynamically independent systems which are coupled to each other via constraints and/or some common, cooperative objective; the distributed MPC schemes developed in this thesis fall into this second category. In such a setting, each system can often only communicate with a certain (possibly small) number of other (neighboring) systems, which requires the synthesis of distributed controllers which only take into account information from such neighboring systems. Again, different stabilizing distributed MPC schemes for these type of systems have been proposed, for example using additional constraints restricting predicted input or state sequences (Dunbar and Murray, 2006; Keviczky et al., 2006; Wang and Ong, 2010), employing some special (sequential) order in which the systems solve their local optimization problems (Grüne and Worthmann, 2012; Richards and How, 2007; Trodden and Richards, 2010, 2013), or using small gain arguments (Franco et al., 2008) and explicit MPC and distributed optimization techniques (Spudić and Baotić, 2013).

The vast majority of existing distributed MPC schemes, including those referenced above, consider the classical control objective of stabilizing the overall system at some (a priori given) setpoint. However, when considering typical cooperative control problems, often different, more general, control objectives are of interest. One particularly important example of such a cooperative control task are consensus and synchronization problems, where the systems have to agree on an a priori unknown state (or output) value or trajectory. These type of problems appear in many applications such as multi-vehicle platooning and formation stabilization, distributed sensor fusion, the synchronization of coupled oscillators and others (see, e.g., Olfati-Saber et al. (2007)). In recent years, significant attention was paid to the analysis and solution of consensus and synchronization problems, see, e.g., (Jadbabaie et al., 2003; Olfati-Saber and Murray, 2004; Olfati-Saber et al., 2007; Wieland, 2010) and the references therein, where, however, control methods different than MPC were used. Only a few distributed MPC schemes have been proposed considering such more general cooperative control tasks. For example, in (Ferrari-Trecate et al., 2009; Zhan and Li, 2013), consensus for single- and double-integrators is considered, exploiting certain geometric properties of the system trajectories resulting from the rather simple dynamics. In (Johansson et al., 2006), for linear systems, the authors calculate an optimal consensus point at each time step by iteratively solving a centralized optimization problem, where this point is used as setpoint in the MPC formulation. A theoretical analysis of this scheme was done by Keviczky and Johansson (2008), where convergence to a common consensus point corresponding to an equilibrium of the system is established. Lee et al. (2011) solve a consensus problem for linear systems using standard methods, and then use the resulting solution as a reference signal in a tracking MPC framework. In a recent work, Gautam et al. (2014) propose a robust distributed MPC scheme for linear time-varying systems to achieve practical consensus. Finally, Dunbar and Caveney (2012) consider stability and string stability in vehicle platoons with a given velocity in a distributed MPC framework.

Furthermore, concerning the field of distributed *economic* MPC (compare Section 1.2.2), there are only few theoretical results available in the literature. (Lee and Angeli, 2011, 2012; Wang, 2011) extend the approach of Stewart et al. (2010), which was developed for setpoint stabilization, to a setting with economic objective function. Driessen et al. (2012) propose a distributed economic MPC scheme for output stabilization in a network of competing systems. Furthermore, Chen et al. (2012) consider the distributed implementation of a Lyapunov-based economic MPC scheme, and Wolf et al. (2012) propose a hierarchical structure where a centralized economic MPC problem is solved on the upper level on a slow time scale, while a distributed controller on the lower level is implemented on a faster time scale to react on disturbances.

In summary, the above discussion shows that only few distributed MPC methods for more general control tasks than setpoint stabilization are available, and hence motivates the need for the development of distributed MPC schemes which are suitable for general cooperative control problems. In this thesis, we develop three such distributed MPC schemes, as will be discussed in more detail in Section 1.3. The first two schemes, presented in Chapter 3, are formulated in the context of stabilizing (tracking) MPC, while the third scheme, developed in Chapter 5, uses the framework of economic MPC.

1.2.2 Economic MPC

Economic MPC is a variant of MPC where some *general* cost function is used within the repeatedly solved optimization problem, which need not be positive definite as is typically assumed in standard stabilizing (tracking) MPC (Angeli et al., 2012; Rawlings et al., 2012). The motivation (and also the naming) for this more general MPC formulation came from the process industry. There, the classical approach is to apply a two-layer control structure, where at the upper layer the so-called real-time optimization (RTO) determines economically optimal setpoints or trajectories for the system, which are then tracked by a controller on the lower level, see, e.g., (Backx et al., 2000; Engell, 2007; Kadam and Marquardt, 2007; Rawlings and Amrit, 2009) and the references therein. However, it was noticed that the performance can possibly be significantly improved if the economic cost function is directly incorporated into the lower level MPC controller (Angeli et al., 2012; Rawlings and Amrit, 2009). In recent years, different economic MPC settings under different assumptions have been studied in the literature, for example with terminal equality constraints (Angeli and Rawlings, 2010; Angeli et al., 2009, 2012; Diehl et al., 2011), with a terminal region constraint (Amrit et al., 2011) and generalized terminal constraints (Fagiano and Teel, 2013; Ferramosca, 2011), or without terminal constraints (Grüne, 2013). Furthermore, Lyapunov-based economic MPC is considered in (Ellis and Christofides, 2014; Heidarinejad, 2012; Heidarinejad et al., 2012), Huang et al. (2011) examine economic MPC for cyclic process operation, and (Maree and Imsland, 2013; Zavala and Flores-Tlacuahuac, 2012) study multi-objective control problems based on economic MPC. Successful applications of economic MPC for various different applications can, e.g., be found in (Chu et al., 2012; Gros, 2013; Hovgaard et al., 2012; Omell and Chmielewski, 2013; Shafiei et al., 2013).

One of the main features of economic MPC is that the resulting closed-loop system is not necessarily convergent, which results from the use of a general (economic) cost function. This observation has several implications. Firstly, it is of interest to obtain bounds on the closed-loop (average) performance. Using suitable terminal constraints, Angeli et al. (2012) and Amrit et al. (2011) establish that the asymptotic average performance is at least as good as the optimal steady-state cost (see Section 2.2 for more details); similar results can be obtained under suitable additional assumptions in a setting without terminal constraints (Grüne, 2013). Secondly, the possibly non-convergent behavior of the closed-loop system raises the question how constraints on average values of input and state variables can be satisfied, see (Angeli et al., 2012) and Section 2.2.2. Thirdly, an important question is under what conditions the optimal operational regime for a system is steady-state operation, and whether in this case the resulting closed-loop system does in fact converge to the optimal steady-state. In this respect, as discussed in more detail in Section 2.2, a certain dissipativity condition has turned out to play a crucial role. Namely, as was shown in (Amrit et al., 2011; Angeli and Rawlings, 2010; Angeli et al., 2012; Diehl et al., 2011; Grüne, 2013), it can be used as a sufficient condition for optimal steady-state operation as well as for closed-loop convergence and stability analysis.

In summary, economic MPC is still a comparatively young field of research, and while important theoretical advances and first successful applications have been accomplished in recent years as discussed above, there are still many important open problems to address. These include (i) a thorough analysis of the mentioned dissipativity property, in particular with respect to its necessity and robustness to parameter changes, (ii) the development and convergence analysis of suitable economic MPC schemes which also take average constraints into account, and (iii) the development of economic MPC schemes using a generalized terminal constraint for which desired closed-loop performance guarantees can be obtained. In Chapter 4, we give answers to these questions, as specified in more detail in the following section.

1.3 Contributions and outline of this thesis

In the following, we specify in more detail the contributions and the outline of this thesis.

Chapter 2 – Background. In this chapter, we provide the necessary background for the remainder of the thesis. In particular, we review basic results in both stabilizing and economic MPC on which the results obtained in the following chapters are based.

Chapter 3 – Distributed MPC for cooperative control. In this chapter, we propose two new distributed MPC schemes for cooperative control. The control objective is translated into the stabilization of some set $\mathbb{X}^*$, which means that the proposed MPC schemes are formulated in the context of stabilizing (tracking) MPC, and the individual performance criterion of each system is consistent with the overall cooperative goal to be achieved. The first method is applicable for cost functions with a general structure, while the second requires a certain separable structure of the cost functions, which allows for a reduced amount of communication between the systems. We discuss and analyze various properties of the two methods and prove constraint satisfaction and achievement of the desired cooperative goal for the resulting closed-loop system. Both distributed MPC schemes are formulated such that they are applicable to general cooperative control tasks. As a specific example, we show how they can be used for consensus and synchronization problems, and we develop a procedure how the required terminal ingredients (terminal cost and terminal region) can be obtained. In summary, the main contributions of this chapter are the following:

- We develop two distributed MPC schemes which guarantee achievement of general cooperative control tasks.

- We show how the schemes can be used for consensus and synchronization problems, and provide a systematic method how to calculate the terminal ingredients.

The results of Chapter 3 are based on (Müller and Allgöwer, 2014a; Müller et al., 2011, 2012b); furthermore, they were also partially summarized in (Grüne et al., 2014, Section 4.3.3).

Chapter 4 – Economic MPC: Dissipativity, average constraints, and self-tuning terminal cost. In this chapter, several contributions in the field of economic MPC are obtained. First, Section 4.1 provides a detailed analysis of a dissipativity condition which plays a crucial role in economic MPC. In particular, we examine whether this dissipativity condition is not only sufficient, but also necessary for optimal steady-state operation of a system. While it turns out that this is not true in the most general case, we show that it is in fact necessary under a (rather mild) controllability condition. This result is also important for convergence and stability analysis in economic MPC, as it guarantees the existence of a storage function under certain dynamic properties of the considered system. Furthermore, we examine robustness of the dissipativity condition with respect to changes in the constraints. This is important as the involved supply rate depends on the optimal steady-state and hence the constraints, and we show with a simple motivating example that arbitrarily small changes in the constraints may result in a loss of dissipativity. We then establish robustness results for both small and large changes in the constraints under some regularity and convexity assumptions, respectively.

In Section 4.2, we develop economic MPC schemes which also take into account average constraints. We first show how *asymptotic* average constraints can be incorporated into a setting with terminal cost and (possibly time-varying) terminal region, and we show how the necessary parameters can be calculated. We then provide a Lyapunov-like analysis to guarantee closed-loop convergence in economic MPC with average constraints under a relaxed dissipativity condition. Finally, we propose an economic MPC scheme which also guarantees satisfaction of *transient* average constraints besides asymptotic average constraints.

In Section 4.3, we develop an economic MPC scheme with self-tuning terminal cost and a generalized terminal constraint. We establish desired closed-loop average performance bounds under general, abstract conditions on the terminal cost. We then propose several specific self-tuning update rules for the terminal weight and show that they satisfy these conditions.

To summarize, the main contributions of this chapter are the following:

- We prove that dissipativity is necessary for optimal steady-state operation under (rather mild) controllability conditions.

- We provide a robustness analysis of dissipativity in economic MPC in the context of changes in the constraint sets.

- We show how satisfaction of both asymptotic and transient average constraints can be guaranteed in economic MPC, using a terminal cost / terminal region framework.

- We provide a Lyapunov-like analysis to study convergence in economic MPC with average constraints under dissipativity.

- We develop an economic MPC framework with self-tuning terminal cost and provide desired closed-loop average performance bounds.

- Throughout the whole chapter, we illustrate the obtained results with various numerical examples, including a chemical reactor application.

The results presented in Chapter 4 are based on (Müller and Allgöwer, 2012; Müller et al., 2013a,b,c,d, 2014b,c,d,f).

Chapter 5 – Distributed economic MPC. In this chapter, we combine ideas from the previous two chapters and develop a distributed economic MPC framework, which is well suited for cooperative control tasks in networks with self-interested systems. The systems cooperate by computing online an (a priori unknown) overall optimal steady-state, while already taking control actions. These are computed by each system via economic MPC, considering its own, individual objective (modeling its self-interest). We study convergence properties of the proposed distributed economic MPC scheme and show that the desired cooperative task is achieved. We illustrate the obtained results with an example of synchronizing several systems with conflicting objective.

The results presented in Chapter 5 are based on (Müller and Allgöwer, 2014b).

Chapter 6 – Conclusions. In this chapter, we summarize the contributions obtained in this thesis and discuss interesting topics for future research.

Appendices A and B provide supplementary material, in particular required stability definitions and a brief introduction to nonlinear programming.

Chapter 2

Background

In this chapter, we provide the necessary background for this thesis. While a general literature review over the field of MPC was already given in Section 1.2, we now discuss in more detail some important results in the frameworks of stabilizing (tracking) and economic MPC.

2.1 Stabilizing MPC

We consider discrete-time nonlinear systems of the form

$$x(t+1) = f(x(t), u(t)), \qquad x(0) = x_0, \tag{2.1}$$

where $f : \mathbb{X} \times \mathbb{U} \to \mathbb{R}^n$, $x(t) \in \mathbb{X} \subseteq \mathbb{R}^n$ and $u(t) \in \mathbb{U} \subseteq \mathbb{R}^m$ are the system state and the control input, respectively, at time $t \in \mathbb{I}_{\geq 0}$, and $x_0 \in \mathbb{X}$ is the initial condition. The system is subject to possibly mixed pointwise-in-time state and input constraints

$$(x(t), u(t)) \in \mathbb{Z} \subseteq \mathbb{X} \times \mathbb{U} \tag{2.2}$$

for all $t \in \mathbb{I}_{\geq 0}$; denote by $\mathbb{Z}_{\mathbb{X}}$ the projection of $\mathbb{Z}$ on $\mathbb{X}$, i.e., $\mathbb{Z}_{\mathbb{X}} := \{x \in \mathbb{X} : \exists u \in \mathbb{U} \text{ s.t. } (x, u) \in \mathbb{Z}\}$. Let S be defined as the set of all feasible state/input equilibrium pairs of system (2.1), i.e.,

$$S := \{(x, u) \in \mathbb{Z} : x = f(x, u)\}, \tag{2.3}$$

which is assumed to be non-empty. The control objective is to stabilize[1] the system (2.1) at a given setpoint x^*, which is an equilibrium point of system (2.1) with corresponding control input u^*, i.e., $(x^*, u^*) \in S$. We impose the following standard assumptions on the system dynamics and the state and input constraint sets.

Assumption 2.1. *The function f is continuous.*

Assumption 2.2. *The set $\mathbb{U}$ is compact, and the set $\mathbb{Z} \subseteq \mathbb{X} \times \mathbb{U}$ is closed.*

Remark 2.1. *While compactness of $\mathbb{U}$ is typically assumed in the MPC literature for technical reasons (in particular in order to guarantee existence of the solution to Problem 2.2 below), note that this assumption could also be relaxed to $\mathbb{U}$ being closed under certain conditions on the cost function ℓ, see (Rawlings and Mayne, 2012) for more details.*

---

[1]See Appendix A.1 for the required background on stability of discrete-time systems.

In order to define the receding horizon control law, at each time $t \in \mathbb{I}_{\geq 0}$ with measured state $x(t)$, the following optimization problem is solved:

Problem 2.2.

$$\underset{\boldsymbol{u}(t)}{minimize} \; J_N(x(t), \boldsymbol{u}(t))$$

subject to

$$x(k+1|t) = f(x(k|t), u(k|t)), \quad k \in \mathbb{I}_{[0,N-1]}, \tag{2.4a}$$
$$x(0|t) = x(t), \tag{2.4b}$$
$$(x(k|t), u(k|t)) \in \mathbb{Z}, \quad k \in \mathbb{I}_{[0,N-1]}, \tag{2.4c}$$
$$x(N|t) \in \mathbb{X}^f, \tag{2.4d}$$

where

$$J_N(x(t), \boldsymbol{u}(t)) := \sum_{k=0}^{N-1} \ell(x(k|t), u(k|t)) + V^f(x(N|t)). \tag{2.5}$$

In this problem, $\boldsymbol{u}(t) := \{u(0|t), \ldots, u(N-1|t)\}$ and $\boldsymbol{x}(t) := \{x(0|t), \ldots, x(N|t)\}$ are input and corresponding state sequences predicted at time t over the prediction horizon $N \in \mathbb{I}_{\geq 0}$. Furthermore, $\mathbb{X}^f \subseteq \mathbb{X}$ is the (closed) terminal region, and $\ell : \mathbb{X} \times \mathbb{U} \to \mathbb{R}$ and $V^f : \mathbb{X}^f \to \mathbb{R}$ are the stage and terminal cost functions, respectively, for which we make the following assumption.

Assumption 2.3. *The stage and terminal cost functions ℓ and V^f are continuous.*

Denote the minimizer of Problem 2.2 by[2] $\boldsymbol{u}^0(t) := \{u^0(0|t), \ldots, u^0(N-1|t)\}$, the corresponding state sequence by $\boldsymbol{x}^0(t) := \{x^0(0|t), \ldots, x^0(N|t)\}$, and by $J_N^0(x(t)) := J_N(x(t), \boldsymbol{u}^0(t))$ the corresponding optimal value function. The model predictive controller is now defined by the following algorithm.

Algorithm 2.3 (Model predictive control for discrete-time systems). *Consider system (2.1). At each time $t \in \mathbb{I}_{\geq 0}$, measure the state $x(t)$, solve Problem 2.2 and apply the control input $u(t) := u^0(0|t)$.*

Algorithm 2.3 results in the closed-loop system

$$x(t+1) = f(x(t), u^0(0|t)), \qquad x(0) = x_0. \tag{2.6}$$

In order to guarantee that x^* is an asymptotically stable equilibrium point for system (2.6), the following conditions are typically imposed on the stage and terminal cost functions ℓ and V^f, respectively, and the terminal region $\mathbb{X}^f$ (see, e.g., Rawlings and Mayne (2009)).

Assumption 2.4. *The stage cost function ℓ satisfies $\ell(x^*, u^*) = 0$ and there exists a function $\alpha_1 \in \mathcal{K}_\infty$ such that $\ell(x, u) \geq \alpha_1(|x - x^*|)$ for all $(x, u) \in \mathbb{Z}$. Furthermore, the terminal cost function V^f satisfies $V^f(x^*) = 0$ and $V^f(x) \geq 0$ for all $x \in \mathbb{X}^f$.*

---

[2]Note that $\boldsymbol{u}^0$ exists due to continuity of ℓ, V^f and f, compactness of $\mathbb{U}$ and closedness of $\mathbb{Z}$ and $\mathbb{X}^f$. Furthermore, for simplicity, we assume that $\boldsymbol{u}^0$ is unique; if this is not the case, just assign a unique constant selection map to select one of the multiple minima. This consideration also applies in all of the following chapters of this thesis.

Assumption 2.5. *The terminal region $\mathbb{X}^f \subseteq \mathbb{X}$ is closed and contains x^* in its interior. Furthermore, there exists a local auxiliary control law $u = \kappa^f(x)$ such that the following is satisfied for all $x \in \mathbb{X}^f$:*

i) $(x, \kappa^f(x)) \in \mathbb{Z}$

ii) $f(x, \kappa^f(x)) \in \mathbb{X}^f$

iii) $V^f(f(x, \kappa^f(x))) - V^f(x) \leq -\ell(x, \kappa^f(x)) + \ell(x^*, u^*)$

Assumption 2.5 means that when the local auxiliary controller is applied to system (2.1), *i)* state and input constraints are satisfied inside the terminal region and *ii)* the terminal region is invariant. Furthermore, in case that Assumption 2.4 is satisfied, condition *iii)* means that V^f serves as a control Lyapunov function inside the terminal region (V^f is positive definite with respect to x^* in this case). Assumptions 2.4 and 2.5 are standard assumptions used in stabilizing (tracking) MPC in a terminal cost / terminal region setting, and we refer the interested reader to (Chen and Allgöwer, 1998; Rawlings and Mayne, 2009) for procedures how V^f, $\mathbb{X}^f$ and κ^f can be calculated such that Assumption 2.5 is satisfied.

Given the above, the following result can be obtained concerning stability of the closed-loop system (2.6). To this end, denote by $\mathbb{X}_N$ the set of all states $x \in \mathbb{X}$ such that Problem 2.2 (with $x(t) = x$) has a solution.

Theorem 2.4. *Consider system (2.1) with $x_0 \in \mathbb{X}_N$ and suppose that Assumptions 2.1–2.5 are satisfied. Then Problem 2.2 is feasible for all $t \in \mathbb{I}_{\geq 0}$. Furthermore, for the closed-loop system (2.6) resulting from application of Algorithm 2.3, the pointwise-in-time state and input constraints (2.2) are satisfied for all $t \in \mathbb{I}_{\geq 0}$, and x^* is an asymptotically stable equilibrium point with region of attraction $\mathbb{X}_N$.*

Theorem 2.4 and its proof can, e.g., be found in (Grüne and Pannek, 2011; Mayne et al., 2000; Rawlings and Mayne, 2009); see also the continuous-time versions in (Chen and Allgöwer, 1998; Findeisen et al., 2003; Fontes, 2001).

2.2 Economic MPC

We now turn our attention to *economic* MPC. As already discussed in Chapter 1, the key difference of economic MPC compared to stabilizing MPC is that Assumption 2.4 is not supposed to hold anymore, i.e., the stage cost function ℓ is not assumed to achieve its minimum at a specific equilibrium but can be an arbitrary cost function, possibly resembling the economics related to the considered system. As a consequence, a particular feature of economic MPC is that the resulting closed-loop system (2.6) may not be convergent. For some typical results in economic MPC, Assumption 2.2 needs to be slightly strengthened and is usually replaced by the following assumption.

Assumption 2.6. *The set $\mathbb{Z} \subseteq \mathbb{X} \times \mathbb{U}$ is compact.*

Note that this is, in general, not a very restrictive assumption. Nevertheless, to be more general, in the remainder of this thesis we state our results without the boundedness assumption on $\mathbb{Z}$ whenever possible, or remark when and how it can be relaxed.

2.2.1 Average performance, optimal steady-state operation and stability

For a given cost function ℓ, we now define the set of optimal state/input equilibrium pairs as

$$S^* := \{(y, w) \in S : \ell(y, w) = \min_{(x,u)\in S} \ell(x, u)\} \tag{2.7}$$

with S as defined in (2.3). Note that S^* is non-empty and well defined, i.e., the minimum in (2.7) exists, if Assumptions 2.3 and 2.6 are satisfied (the latter implies that S is compact). On the other hand, in case that S is not compact (which can happen if only Assumption 2.2 instead of Assumption 2.6 is imposed), we still assume that S^* is non-empty and well defined, i.e., the minimum in (2.7) exists. Furthermore, note that in general S^* need not be a singleton, i.e., there does not necessarily exist a unique optimal feasible state/input equilibrium pair (x^*, u^*). In the following, by (x^*, u^*) we denote an arbitrary element of the set S^*. Given the above, the following result concerning the asymptotic average performance of the closed-loop system (2.6) has been established in (Amrit et al., 2011; Angeli et al., 2012).

Theorem 2.5. *Suppose that Assumptions 2.1, 2.3, 2.5 (with $(x^*, u^*) \in S^*$) and 2.6 are satisfied and that $x_0 \in \mathbb{X}_N$. Then the asymptotic average performance of the closed-loop system (2.6) resulting from application of Algorithm 2.3 is at least as good as the optimal equilibrium cost, i.e., $\limsup_{T\to\infty} \frac{\sum_{t=0}^{T} \ell(x(t),u(t))}{T+1} \leq \ell(x^*, u^*)$ holds along the solution of the closed-loop system (2.6).*

A proof of Theorem 2.5 was first obtained for the case of a terminal equality constraint, i.e., $\mathbb{X}^f = \{x^*\}$ and $V^f \equiv 0$ and can be found in (Angeli et al., 2012). Under the given hypotheses, i.e., with a nontrivial terminal region $\mathbb{X}^f$, Theorem 2.5 was then established by Amrit et al. (2011). In both cases, the proof proceeds by showing that along the resulting closed-loop system (2.6), we have $J_N^0(x(t+1)) - J_N^0(x(t)) \leq \ell(x^*, u^*) - \ell(x(t), u(t))$, from where it then follows that

$$0 \leq \liminf_{T\to\infty} \frac{J_N^0(x(T+1)) - J_N^0(x_0)}{T+1} \leq \ell(x^*, u^*) - \limsup_{T\to\infty} \frac{\sum_{t=0}^{T} \ell(x(t), u(t))}{T+1} \tag{2.8}$$

as claimed. Note that the first inequality in (2.8) follows from Assumptions 2.3 and 2.6, i.e., continuity of ℓ and V^f and compactness of $\mathbb{Z}$. Namely, $J_N^0(x)$ is bounded for all $x \in \mathbb{X}_N$ in this case, from where the first inequality in (2.8) follows (in fact, with equality). On the other hand, one could also allow for an unbounded constraint set $\mathbb{Z}$, i.e., replace Assumption 2.6 in Theorem 2.5 with Assumption 2.2, if one additionally directly assumes that the first inequality in (2.8) is satisfied. A sufficient condition for this is, e.g., that both ℓ and V^f are bounded from below on $\mathbb{Z}$ and $\mathbb{X}^f$, respectively.

Furthermore, in (Amrit et al., 2011), under the assumptions that f and ℓ are twice continuously differentiable, the linearization of system (2.1) at (x^*, u^*) is stabilizable and (x^*, u^*) is in the interior of $\mathbb{Z}$, a method was developed how $\mathbb{X}^f$, V^f and κ^f can be computed such that Assumption 2.5 is satisfied. Note that this procedure is different from those available in the setting of stabilizing MPC (see Section 2.1), as Assumption 2.4 is no longer assumed to hold, and the resulting terminal cost V^f is in general not positive definite with respect to x^*.

As already discussed above, due to the fact that Assumption 2.4 is not supposed to hold, non-constant trajectories of system (2.1) might exist that outperform the optimal steady-state. In order to classify in a rigorous way for which systems steady-state operation is optimal, we use the following definition from[3] (Angeli and Rawlings, 2010; Angeli et al., 2012).

Definition 2.6 (Optimal operation at steady-state). *The system* (2.1) *is* optimally operated at steady-state *with respect to the cost function ℓ, if for each solution satisfying $(x(t), u(t)) \in \mathbb{Z}$ for all $t \in \mathbb{I}_{\geq 0}$ the following holds:*

$$\liminf_{T \to \infty} \frac{\sum_{t=0}^{T} \ell(x(t), u(t))}{T + 1} \geq \ell(x^*, u^*), \tag{2.9}$$

where $(x^*, u^*) \in S^*$. The system (2.1) *is* suboptimally operated off steady-state, *if in addition at least one of the following two conditions holds:*

$$\liminf_{T \to \infty} \frac{\sum_{t=0}^{T} \ell(x(t), u(t))}{T + 1} > \ell(x^*, u^*) \tag{2.10a}$$

$$\liminf_{t \to \infty} |x(t) - x^*| = 0 \tag{2.10b}$$

The definition of optimal operation at steady-state is such that no feasible solution to system (2.1) leads to an average performance (measured in terms of the stage cost ℓ) which is better than operation of the system at the optimal steady-state x^*. Note that the question of optimal steady-state operation is related to the field of optimal periodic control. Namely, in the classical literature on optimal periodic control, various sufficient conditions (usually based on first or second variation analysis) are available to determine whether a periodic trajectory exists which yields a better performance than the optimal steady-state solution (see, e.g., Bittanti and Guardabassi (1986); Bittanti et al. (1973); Gilbert (1977)). On the other hand, a certain *dissipativity* condition was recently used to study optimal steady-state operation of a system (Angeli and Rawlings, 2010; Angeli et al., 2012). The notion of dissipativity was introduced in (Willems, 1972), see also (Byrnes and Lin, 1994) for a discrete time version; we adapt it here to our setting including state and input constraints, similar to (Angeli et al., 2012). To this end, for a set $\mathbb{W} \subseteq \mathbb{Z}$, denote by $\mathbb{W}_{\mathbb{X}}$ the projection of $\mathbb{W}$ on $\mathbb{X}$, i.e., $\mathbb{W}_{\mathbb{X}} := \{x \in \mathbb{X} : \exists u \in \mathbb{U} \text{ s.t. } (x, u) \in \mathbb{W}\}$.

Definition 2.7 (Dissipativity). *The system* (2.1) *is* dissipative *on a set $\mathbb{W} \subseteq \mathbb{Z}$ with respect to the supply rate $s : \mathbb{W} \to \mathbb{R}$ if there exists a storage function $\lambda : \mathbb{W}_{\mathbb{X}} \to \mathbb{R}$, which is bounded from below on $\mathbb{W}_{\mathbb{X}}$, such that the following inequality is satisfied for all $(x, u) \in \mathbb{W}$:*

$$\lambda(f(x, u)) - \lambda(x) \leq s(x, u). \tag{2.11}$$

If, in addition, for some $\rho \in \mathcal{K}_{\infty}$ it holds that for all $(x, u) \in \mathbb{W}$

$$\lambda(f(x, u)) - \lambda(x) \leq -\rho(|x - x^*|) + s(x, u), \tag{2.12}$$

then system (2.1) *is* strictly dissipative *on $\mathbb{W}$.*

---

[3]In (Angeli et al., 2012), optimal steady-state operation was defined using the equation $Av[\ell(x, u)] \subseteq [\ell(x^*, u^*), \infty)$ instead of (2.9). While these two statements are in fact equivalent, in this thesis we use (2.9) in order to be able to also consider a possibly unbounded constraint set $\mathbb{Z}$, in which case $Av[\ell(x, u)]$ is not necessarily well defined as ℓ might not be bounded.

Remark 2.8. *Note that in the original definition (Byrnes and Lin, 1994; Willems, 1972), the storage function λ is required to be nonnegative. In accordance with (Amrit et al., 2011; Angeli et al., 2012; Grüne, 2013), we do not impose this assumption here but only require that it is bounded from below. In fact, one could then also just add a constant to make λ nonnegative on $\mathbb{W}_{\mathbb{X}}$.*

Now define the set $\mathbb{Z}^0$ as the largest "forward invariant" set contained in $\mathbb{Z}$, i.e., the set which contains all elements in $\mathbb{Z}$ which are part of a feasible state/input sequence pair:

$$\mathbb{Z}^0 := \Big\{ (x,u) \in \mathbb{Z} : \exists \boldsymbol{v} : \mathbb{I}_{\geq 0} \to \mathbb{U} \text{ s.t. } (z(0),v(0)) = (x,u), \; z(t+1) = f(z(t),v(t)),$$

$$(z(t),v(t)) \in \mathbb{Z} \; \forall t \in \mathbb{I}_{\geq 0}, \Big\} \subseteq \mathbb{Z}. \tag{2.13}$$

Denote by $\mathbb{X}^0$ the projection of $\mathbb{Z}^0$ on $\mathbb{X}$, i.e., $\mathbb{X}^0 := \{x \in \mathbb{X} : \exists u \in \mathbb{U} \text{ s.t. } (x,u) \in \mathbb{Z}^0\}$. Note that both $\mathbb{Z}^0$ and $\mathbb{X}^0$ are compact in case that Assumptions 2.1 and 2.6 are satisfied, i.e., f is continuous and $\mathbb{Z}$ is compact. We can now state the following result, which connects dissipativity of system (2.1) to optimal steady-state operation and can be found in Angeli et al. (2012).

Theorem 2.9. *Suppose that system (2.1) is dissipative (strictly dissipative) on $\mathbb{Z}^0$ with respect to the supply rate $s(x,u) = \ell(x,u) - \ell(x^*,u^*)$. Then the system (2.1) is optimally operated at steady-state (suboptimally operated off steady-state).*

Furthermore, besides classifying optimal steady-state operation, (strict) dissipativity of system (2.1) is also crucial for establishing stability of the closed-loop system (2.6) in an economic MPC setting, as shown in the following result from (Amrit et al., 2011; Angeli et al., 2012).

Theorem 2.10. *Suppose that Assumptions 2.1, 2.3, 2.5 and 2.6 are satisfied, that system (2.1) is strictly dissipative on $\mathbb{Z}^0$ with respect to the supply rate $s(x,u) = \ell(x,u) - \ell(x^*,u^*)$, and the corresponding storage function λ is continuous at x^*. Then, x^* is an asymptotically stable equilibrium point of the closed-loop system (2.6) resulting from application of Algorithm 2.3, with $\mathbb{X}_N$ being the region of attraction.*

Again, Theorem 2.10 was first obtained by Angeli et al. (2012) for the case of a terminal equality constraint, i.e., $\mathbb{X}^f = \{x^*\}$ and $V^f \equiv 0$; note that in this case an additional weak controllability assumption is necessary (see Assumption 2 in Angeli et al. (2012)). Under the given hypotheses, i.e., with a nontrivial terminal region $\mathbb{X}^f$, Theorem 2.10 was then established by[4] Amrit et al. (2011). Furthermore, the dissipativity condition in Theorem 2.10 is also crucial for convergence and stability analysis in an economic MPC setting without terminal constraints, see (Grüne, 2013).

Remark 2.11. *A close inspection of the proof of Theorem 2.10 (see Amrit et al. (2011)) reveals that one can replace Assumption 2.6 by Assumption 2.2, i.e., allow for an unbounded*

---

[4]In Amrit et al. (2011), the storage function λ was assumed to be continuous on its whole domain. However, continuity at x^* as in Theorem 2.10 is in fact enough to establish closed-loop asymptotic stability as in Definition A.1 in Appendix A. On the other hand, for asymptotic stability according to the slightly stronger Definition A.2, one would need the (very mild) additional assumption that λ is locally bounded on $\mathbb{X}^0$ (which clearly is, e.g., implied by λ being continuous on $\mathbb{X}^0$).

constraint set $\mathbb{Z}$, *under the (minor) additional assumption that* V^f *is bounded from below on* $\mathbb{X}^f$. *Note that this is, e.g., satisfied if the terminal region* $\mathbb{X}^f$ *is compact (due to continuity of* V^f).

As a conclusion of the above results, one can see that dissipativity plays a crucial role in economic MPC. As already stated in Section 1.3, one of the contributions of this thesis (see Section 4.1) is to thoroughly investigate the employed dissipativity condition. In particular, we examine whether it is not only sufficient, but also necessary for optimal steady-state operation (Section 4.1.1), and establish robustness properties in case of changing constraint sets (Section 4.1.2).

Remark 2.12. *When actually searching for a storage function* λ *for a given system in order to establish dissipativity, one might rather consider the possibly larger set* $\mathbb{Z}$ *instead of* $\mathbb{Z}^0$ *(as in Theorems 2.9 and 2.10) and establish dissipativity there, as* $\mathbb{Z}^0$ *might in general be difficult to compute. Furthermore, while for a linear system with convex constraints and cost function, methods exist how* λ *can be determined (Damm et al., 2014; Diehl et al., 2011), for a general nonlinear system with nonconvex constraints and cost function this can be a difficult task and, to the best of the author's knowledge, no systematic procedure exists. In this respect, the results of Section 4.1.1 are of particular interest as existence of a storage function* λ *is sufficient for stability analysis (see Theorem 2.10) of the closed-loop system, but* λ *does not have to be known; the same is true in economic MPC without terminal constraints, see (Grüne, 2013).*

2.2.2 Average constraints

In economic MPC, besides the considered pointwise-in-time state and input constraints, also constraints on asymptotic average values of states and inputs are of interest. Namely, as discussed in Section 2.1, standard tracking MPC is designed such that the closed-loop system converges to some equilibrium setpoint, which means that any asymptotic average of state and input variables is determined by their value at this equilibrium. Therefore, asymptotic average constraints are not of further interest online, i.e., during application of Algorithm 2.3, but have to be taken into account offline when determining the setpoint to be stabilized. On the other hand, as discussed above, one of the key features of economic MPC is that the closed-loop system may not be convergent. Hence it is natural to also consider constraints on average values of input and state variables, which now have to be taken into account online, i.e., within the MPC algorithm, see (Angeli et al., 2012). To be precise, as in (Angeli et al., 2012), for some (auxiliary) output function $h : \mathbb{X} \times \mathbb{U} \to \mathbb{R}^p$ and some $\mathbb{Y} \subseteq \mathbb{R}^p$, the asymptotic average constraints which have to hold along solutions of system (2.1) are given as[5]

$$Av[h(\boldsymbol{x}, \boldsymbol{u})] \subseteq \mathbb{Y}. \tag{2.14}$$

We make the following assumption on h and $\mathbb{Y}$.

Assumption 2.7. *The function* $h : \mathbb{X} \times \mathbb{U} \to \mathbb{R}^p$ *is continuous and the set* $\mathbb{Y} \subseteq \mathbb{R}^p$ *is closed and convex.*

---

[5]With a slight abuse of notation, in the following we denote by $h(\boldsymbol{x}, \boldsymbol{u})$ the output sequence $\{h(x(0), u(0)), h(x(1), u(1)), \dots\}$ for some input and corresponding state sequences $\boldsymbol{u}$ and $\boldsymbol{x}$, respectively, of system (2.1).

Analogous to (2.3) and (2.7), one can define the set of feasible, respectively, optimal state/input equilibrium pairs with additional average constraints as

$$S_{\mathrm{av}} := \{(x, u) \in \mathbb{Z} : x = f(x, u), h(x, u) \in \mathbb{Y}\}, \tag{2.15}$$

$$S_{\mathrm{av}}^* := \{(y, w) \in S_{\mathrm{av}} : \ell(y, w) = \min_{(x, u) \in S_{\mathrm{av}}} \ell(x, u)\}. \tag{2.16}$$

With this, optimal steady-state operation of system (2.1) with additional asymptotic average constraints can be defined analogous to Definition 2.6, where now $(x^*, u^*) \in S_{\mathrm{av}}^*$, and (2.9)–(2.10) have to hold for each solution of system (2.1) satisfying $(x(t), u(t)) \in \mathbb{Z}$ for all $t \in \mathbb{I}_{\geq 0}$ and additionally $Av[h(\boldsymbol{x}, \boldsymbol{u})] \subseteq \mathbb{Y}$. Furthermore, analogous to (2.13) we define the set $\mathbb{Z}_{\mathrm{av}}^0$ as

$$\mathbb{Z}_{\mathrm{av}}^0 := \Big\{(x, u) \in \mathbb{Z} : \exists \boldsymbol{v} : \mathbb{I}_{\geq 0} \to \mathbb{U} \text{ s.t. } (z(0), v(0)) = (x, u), z(t+1) = f(z(t), v(t)),$$

$$(z(t), v(t)) \in \mathbb{Z} \ \forall t \in \mathbb{I}_{\geq 0}, \ Av[h(\boldsymbol{z}, \boldsymbol{v})] \subseteq \mathbb{Y}\Big\} \subseteq \mathbb{Z}^0 \subseteq \mathbb{Z}, \tag{2.17}$$

and denote by $\mathbb{X}_{\mathrm{av}}^0$ the projection of $\mathbb{Z}_{\mathrm{av}}^0$ on $\mathbb{X}$, i.e., $\mathbb{X}_{\mathrm{av}}^0 := \{x \in \mathbb{X} : \exists u \in \mathbb{U} \text{ s.t. } (x, u) \in \mathbb{Z}_{\mathrm{av}}^0\}$.

In (Angeli et al., 2012), it was shown that under Assumption 2.7 and for the considered setting with additional asymptotic average constraints,

a) if a terminal equality constraint is used, i.e., $\mathbb{X}^f = \{x^*\}$ and $V^f \equiv 0$, and if Problem 2.2 is appropriately modified by an additional constraint (see Section 4.2 for more details), then both the average constraints (2.14) are satisfied and the asymptotic average performance result of Theorem 2.5 is valid for the closed-loop system (2.6),

b) for $\mathbb{Y} = \{y \in \mathbb{R}^p : Ay \leq b\}$ with $A \in \mathbb{R}^{n_y \times p}$ and $b \in \mathbb{R}^{n_y}$ for some $n_y \in \mathbb{I}_{\geq 1}$, the results of Theorem 2.9 are still valid with $s(x, u)$ replaced by $\bar{s}(x, u) = \ell(x, u) - \ell(x^*, u^*) + \bar{\lambda}^T h(x, u)$ for some (arbitrary) $\bar{\lambda} \in \mathbb{R}_{\geq 0}^p$, and $\mathbb{Z}^0$ replaced by $\mathbb{Z}_{\mathrm{av}}^0$.

In this thesis, we consider economic MPC with average constraints in Section 4.2. As already stated in Section 1.3, our contributions are (i) to develop an economic MPC framework with additional asymptotic average constraints and a nontrivial terminal region instead of a terminal equality constraint (see Section 4.2.1); (ii) to show that the closed-loop system asymptotically converges to x^* under an appropriate dissipativity condition (see Section 4.2.2); (iii) to develop an economic MPC framework which takes into account *transient* instead of asymptotic average constraints, i.e., constraints on state and input variables averaged over some finite time period (see Section 4.2.3). We believe that the concept of (both transient and asymptotic) average constraints is of interest in various applications such as the operation of a chemical reactor, where e.g. the average amount of inflow or the average heat flux through the reactor wall must not exceed a certain value. Such an example is considered in Section 4.2.4.

2.3 Summary

In this chapter, we have recalled basic facts in stabilizing and economic MPC, which are the background for the remainder of this thesis. In particular, the results presented in Chapter 3 are formulated within the former framework, while those in Chapters 4 and 5 fall into the latter category. To summarize, we have shown that under a suitable set of assumptions, model predictive control results in an asymptotically stable closed-loop system in case that the stage cost ℓ is positive definite with respect to some setpoint (stabilizing MPC); if ℓ does not satisfy this condition, the resulting closed-loop system might not be convergent (economic MPC). Nevertheless, in the latter case, statements on the asymptotic average performance can be obtained, as well as on optimal steady-state operation and also stability if a particular dissipativity condition is satisfied.

Chapter 3

Distributed MPC for cooperative control

In this chapter, we present a distributed MPC framework suitable for solving cooperative control problems. As motivated and discussed in Sections 1.2.1 and 1.3, in contrast to most of the available results in the literature, the proposed distributed MPC schemes can be used for general cooperative control tasks and not only for the stabilization of some (a priori given) setpoint. After introducing the problem setup in Section 3.1, we present two distributed MPC schemes, one using general cost functions and one using cost functions of a certain separable structure (see Section 3.2). These distributed MPC schemes are formulated in a rather abstract and general way, in order to be suitable for a variety of cooperative control problems. In Section 3.3, we then exemplarily show how the developed methods can be applied to consensus and synchronization problems.

The results presented in this chapter are based on (Müller and Allgöwer, 2014a; Müller et al., 2011, 2012b); furthermore, they were also partially summarized in (Grüne et al., 2014, Section 4.3.3).

3.1 Problem setup

Consider a network of $P \in \mathbb{I}_{\geq 0}$ discrete-time nonlinear systems of the form (2.1), i.e.,

$$x_i(t+1) = f_i(x_i(t), u_i(t)), \qquad x_i(0) = x_{i0}, \tag{3.1}$$

with $i \in \mathbb{I}_{[1,P]}$, where $f_i : \mathbb{X}_i \times \mathbb{U}_i \to \mathbb{R}^{n_i}$, $x_i(t) \in \mathbb{X}_i \subseteq \mathbb{R}^{n_i}$ and $u_i(t) \in \mathbb{U}_i \subseteq \mathbb{R}^{m_i}$ are system i's state and control input, respectively, at time $t \in \mathbb{I}_{\geq 0}$, and $x_{i0} \in \mathbb{X}_i$ is its initial condition. Each system is subject to possibly mixed pointwise-in-time local state and input constraints

$$(x_i(t), u_i(t)) \in \mathbb{Z}_i \subseteq \mathbb{X}_i \times \mathbb{U}_i \tag{3.2}$$

for all $t \in \mathbb{I}_{\geq 0}$. As in Chapter 2, we make the following standard assumptions on the system dynamics and the state and input constraint sets.

Assumption 3.1. *The function f_i is continuous for all $i \in \mathbb{I}_{[1,P]}$.*

Assumption 3.2. *For all $i \in \mathbb{I}_{[1,P]}$, the set $\mathbb{U}_i$ is compact, and the set $\mathbb{Z}_i \subseteq \mathbb{X}_i \times \mathbb{U}_i$ is closed.*

Denote by $x(t)$ and $u(t)$ the state and input of the overall system at time t, i.e., $x(t) := [x_1(t)^T \ \ldots \ x_P(t)^T]^T \in \mathbb{R}^{n_{\mathrm{ov}}}$ (with $n_{\mathrm{ov}} = \sum_{i=1}^{P} n_i$) and $u(t) := [u_1(t)^T \ \ldots \ u_P(t)^T]^T \in \mathbb{R}^{m_{\mathrm{ov}}}$ (with $m_{\mathrm{ov}} = \sum_{i=1}^{P} m_i$), and let $\mathbb{X} := \mathbb{X}_1 \times \cdots \times \mathbb{X}_P$, $\mathbb{U} := \mathbb{U}_1 \times \cdots \times \mathbb{U}_P$ and $\mathbb{Z} := \mathbb{Z}_1 \times \cdots \times \mathbb{Z}_P$. Furthermore, denote by $f : \mathbb{X} \times \mathbb{U} \to \mathbb{R}^{n_{\mathrm{ov}}}$ the overall system dynamics such that $x(t+1) = f(x(t), u(t))$.

While the systems (3.1) are dynamically decoupled (i.e., the system dynamics (3.1) of each system is independent of states and inputs of other systems), they are coupled with each other via common constraints and a common objective. In particular, we assume that the overall system is subject to $R \in \mathbb{I}_{\geq 0}$ pointwise-in-time coupling constraints of the form

$$c_r(x(t)) \in \mathbb{C}_r \tag{3.3}$$

for all $t \in \mathbb{I}_{\geq 0}$, where $c_r : \mathbb{X} \to \mathbb{R}^{l_r}$ and $\mathbb{C}_r \subseteq \mathbb{R}^{l_r}$ with $r \in \mathbb{I}_{[1,R]}$ are the R coupling outputs and corresponding coupling constraint sets, respectively. Such coupling constraints could, for example, represent collision avoidance or connectivity maintenance constraints. For each system $i \in \mathbb{I}_{[1,P]}$, define by $\mathcal{R}_i \subseteq \mathbb{I}_{[1,R]}$ the set of all coupling constraints which involve system i, i.e., such that x_i appears explicitly in[1] c_r. Furthermore, for each system $i \in \mathbb{I}_{[1,P]}$, we define a system $j \neq i$ to be a neighbor of system i if the two systems are subject to coupled constraints, i.e., $\mathcal{R}_i \cap \mathcal{R}_j \neq \emptyset$, or share a coupled objective function according to the cooperative goal to be achieved (see below). This means that each of the systems can be identified with a vertex of a graph $\mathcal{G} = (\mathcal{V}, \mathcal{E})$, where $\mathcal{V} = \{v_1, ..., v_P\}$ is the set of vertices, and the set of edges $\mathcal{E} \subseteq \{(v_i, v_j) \in \mathcal{V} \times \mathcal{V} | i \neq j\}$ describes the interconnection topology of the systems. Let $\mathcal{N}_i := \{j | (v_i, v_j) \in \mathcal{E}\}$ be the set of indices of the neighbors of system i, and $d_i := |\mathcal{N}_i|$ its cardinality. According to the above definition of a neighbor, system i is a neighbor of system j if and only if also system j is a neighbor of system i, as coupling constraints and a coupled objective function affect both systems. This means that we model the graph $\mathcal{G}$ to be undirected, i.e., for all $i, j \in \mathbb{I}_{[1,P]}$, $(v_i, v_j) \in \mathcal{E}$ if and only if also $(v_j, v_i) \in \mathcal{E}$. In order to be able to achieve the common objective and to satisfy the coupling constraints, in the following we assume that each system can communicate with each of its neighbors. In particular, this means that the systems can exchange information about predicted sequences with their neighbors, which will be described in more detail later.

As stated above, our control objective is the achievement of some common, cooperative objective, which we translate into the stabilization of some closed set $\mathbb{X}^* \subseteq \mathbb{X}$. This rather general problem formulation includes as two special cases the situations where *(i)* $\mathbb{X}^*$ is a prespecified setpoint, and *(ii)* $\mathbb{X}^*$ is the consensus subspace, i.e., $\mathbb{X}^* = \{x \in \mathbb{R}^{n_{\mathrm{ov}}} : x_1 = x_2 = \cdots = x_P\}$ (in case that the dimension of the systems is the same, i.e., $n_1 = \cdots = n_P$). The latter case will be treated in more detail in Section 3.3.

In order to define the receding horizon control law, at each time $t \in \mathbb{I}_{\geq 0}$ with measured state $x_i(t)$, we associate to each system $i \in \mathbb{I}_{[1,P]}$ the following optimization problem, whose components will be explained in more detail below:

Problem 3.1.

$$\underset{\boldsymbol{u}_i(t)}{minimize} \ J_{i,N}(x_i(t), \tilde{\boldsymbol{x}}_{-i}(t), \boldsymbol{u}_i(t))$$

---

[1]If the functions c_r are continuously differentiable, $\mathcal{R}_i$ is given by $\mathcal{R}_i := \{r \in \mathbb{I}_{[1,R]} : \partial c_r / \partial x_i \not\equiv 0\}$.

subject to

$$x_i(k+1|t) = f_i(x_i(k|t), u_i(k|t)), \quad k \in \mathbb{I}_{[0,N-1]}, \tag{3.4a}$$

$$x_i(0|t) = x_i(t), \tag{3.4b}$$

$$(x_i(k|t), u_i(k|t)) \in \mathbb{Z}_i, \quad k \in \mathbb{I}_{[0,N-1]}, \tag{3.4c}$$

$$c_r(x_i(k|t), \tilde{x}_{-i}(k|t)) \in \mathbb{C}_r, \quad k \in \mathbb{I}_{[1,N]}, \ r \in \mathcal{R}_i, \tag{3.4d}$$

where $J_{i,N}(x_i(t), \tilde{\boldsymbol{x}}_{-i}(t), \boldsymbol{u}_i(t)) := \sum_{k=0}^{N-1} \ell_i(x_i(k|t), \tilde{x}_{-i}(k|t), u_i(k|t)) + V_i^f(x_i(N|t), \tilde{x}_{-i}(N|t)).$

In this problem, $\boldsymbol{u}_i(t) := \{u_i(0|t), \ldots, u_i(N-1|t)\}$ and $\boldsymbol{x}_i(t) := \{x_i(0|t), \ldots, x_i(N|t)\}$ are input and corresponding state sequences predicted at time t over the prediction horizon $N \in \mathbb{I}_{\geq 0}$, and $\tilde{\boldsymbol{x}}_{-i}(t)$ consists of the planned state sequences of system i's neighbors, i.e.,

$$\tilde{\boldsymbol{x}}_{-i}(t) = \{\tilde{x}_{-i}(0|t), \ldots, \tilde{x}_{-i}(N|t)\} := \begin{bmatrix} \tilde{\boldsymbol{x}}_{i_1}(t) \\ \vdots \\ \tilde{\boldsymbol{x}}_{i_{d_i}}(t) \end{bmatrix} = \begin{bmatrix} \{\tilde{x}_{i_1}(0|t), \ldots, \tilde{x}_{i_1}(N|t)\} \\ \vdots \\ \{\tilde{x}_{i_{d_i}}(0|t), \ldots, \tilde{x}_{i_{d_i}}(N|t)\} \end{bmatrix}, \tag{3.5}$$

where $\{i_1, \ldots, i_{d_i}\}$ is an ordered sequence of the elements of the set $\mathcal{N}_i$, i.e. $i_1 < \cdots < i_{d_i}$. While system i solves Problem 3.1, the planned state sequences of its neighbors, $\tilde{\boldsymbol{x}}_{-i}$, are held as constant parameters. We will specify later how $\tilde{\boldsymbol{x}}_{-i}$ is exactly defined, i.e., which state sequences system i assumes for its neighbors. The functions $\ell_i : \mathbb{X}_i \times \mathbb{X}_{i_1} \times \cdots \times \mathbb{X}_{i_{d_i}} \times \mathbb{U}_i \to \mathbb{R}$ and $V_i^f : \mathbb{X}_i \times \mathbb{X}_{i_1} \times \cdots \times \mathbb{X}_{i_{d_i}} \to \mathbb{R}$ are the stage and terminal cost functions which can depend on the neighboring states $x_{-i} := [x_{i_1}^T \ \ldots \ x_{i_{d_i}}^T]^T$, in accordance with the common cooperative goal to be achieved. Finally, we note that with a slight abuse of notation, for each system $i \in \mathbb{I}_{[1,P]}$ and each $r \in \mathcal{R}_i$, we write c_r in (3.4d) as a function of x_i and x_{-i}, whereas in (3.3) it was defined as a function of the overall state x. This can be done as for each system $i \in \mathbb{I}_{[1,P]}$ and each $r \in \mathcal{R}_i$, by definition the coupling output c_r does not depend on systems $j \notin \mathcal{N}_i$.

Remark 3.2. *For clarity of presentation, in this chapter we consider coupling constraints (3.3) involving only the system states x_i; nevertheless, in a similar way, also coupling constraints involving both the system states x_i and the system inputs u_i can be treated. Similar considerations also apply to the stage cost functions ℓ_i, which instead of only depending on neighboring states x_{-i} could also depend on neighboring inputs.* ☐

Now similar to Section 2.1, we impose the following assumptions on the functions ℓ_i, V_i^f, and c_r, and the sets $\mathbb{C}_r$. To this end, let $V^f(x) := \sum_{i=1}^{P} V_i^f(x_i, x_{-i})$.

Assumption 3.3. *The stage and terminal cost functions ℓ_i and V_i^f, respectively, are continuous for all $i \in \mathbb{I}_{[1,P]}$. Furthermore, for all $r \in \mathbb{I}_{[1,R]}$, the function c_r is continuous and the set $\mathbb{C}_r$ is closed.*

Assumption 3.4. *There exists a function $\alpha_1 \in \mathcal{K}_\infty$ such that $\sum_{i=1}^{P} \ell_i(x_i, x_{-i}, u_i) \geq \alpha_1(|x|_{\mathbb{X}^*})$ for all $(x,u) \in \mathbb{Z}$.*

Next, similar to Assumption 2.5, we make the following assumption about the existence and properties of a terminal region $\mathbb{X}^f$.

Assumption 3.5. *There exists a closed terminal region $\mathbb{X}^f \subseteq \mathbb{X}$ and a $\delta^f > 0$ such that $\{x \in \mathbb{X} : |x|_{\mathbb{X}^*} \leq \delta^f\} \subseteq \mathbb{X}^f$, and for each system $i \in \mathbb{I}_{[1,P]}$ a local auxiliary control law $u_i = \kappa_i^f(x_i, x_{-i})$, such that the following holds for all $x \in \mathbb{X}^f$:*

   i) $(x_i, \kappa_i^f(x_i, x_{-i})) \in \mathbb{Z}_i$, for all $i \in \mathbb{I}_{[1,P]}$

   ii) $f(x, \kappa^f(x)) \in \mathbb{X}^f$, where $\kappa^f := [(\kappa_1^f)^T, \ldots, (\kappa_P^f)^T]^T$

   iii) $c_r(x) \in \mathbb{C}_r$, for all $r \in \mathbb{I}_{[1,R]}$

   iv) $V^f(f(x, \kappa^f(x))) - V^f(x) \leq -\sum_{i=1}^{P} \ell_i(x_i, x_{-i}, \kappa_i^f(x_i, x_{-i}))$

If the set $\mathbb{X}^*$ is compact and contained in the interior of $\mathbb{X}$, then requiring existence of a δ^f as in Assumption 3.5 is equivalent to requiring that $\mathbb{X}^*$ is contained in the interior of $\mathbb{X}^f$ (compare Assumption 2.5, where this was done for the case of $\mathbb{X}^* = \{x^*\}$). Furthermore, note that for each system i, the local auxiliary control law κ_i^f is only allowed to depend on neighboring system states x_{-i}. On the other hand, Assumption 3.5 uses a "centralized" terminal region $\mathbb{X}^f$ for the overall system. We use such a centralized terminal region as in typical distributed control tasks such as consensus problems (see Section 3.3), the decrease condition $iv)$ can in general only be ensured for V^f, i.e., the sum of all terminal cost functions, but not necessarily for each single terminal cost function V_i^f, and also the invariance condition $ii)$ can in general only be ensured for a centralized terminal region, but not necessarily for decoupled ones. As in Chapter 2, we require that the overall system state at the end of the prediction horizon, $x(N|t) := [x_1(N|t)^T \ \ldots \ x_P(N|t)^T]^T$, lies inside the terminal region, i.e.,

$$x(N|t) \in \mathbb{X}^f. \tag{3.6}$$

However, in contrast to the centralized setting in Chapter 2, constraint (3.6) cannot directly be included into Problem 3.1 as an additional terminal constraint, as system i does not have any information about non-neighboring systems. In Section 3.2, we present two distributed MPC schemes where the centralized terminal constraint (3.6) can be implemented in a distributed fashion, as required. To this end, for the remainder of this chapter we choose $\mathbb{X}^f$ to be a sublevel set of the sum of the terminal cost functions, i.e., we make the following assumption.

Assumption 3.6. *The terminal region $\mathbb{X}^f$ is of the form $\mathbb{X}^f := \{x \in \mathbb{R}^{n_{ov}} : V^f(x) \leq a\}$ for some $a > 0$.*

Note that if Assumption 3.6 is satisfied, the invariance condition $ii)$ of Assumption 3.5 is implied by condition $iv)$. Finally, we impose the following assumption on the sum V^f of the terminal cost functions.

Assumption 3.7. *There exists a function $\alpha_2 \in \mathcal{K}_\infty$ such that $0 \leq V^f(x) \leq \alpha_2(|x|_{\mathbb{X}^*})$ for all $x \in \mathbb{X}^f$.*

Remark 3.3. *In case that the set $\mathbb{X}^*$ is compact, existence of a $\mathcal{K}_\infty$-function α_2 satisfying Assumption 3.7 is implied by V^f being continuous and $V^f(x) = 0$ for all $x \in \mathbb{X}^*$, even in case that $\mathbb{X}^f$ is unbounded (see Proposition 11 in Rawlings and Mayne (2012)). This was, e.g., the case in Section 2.1, where the control objective was to stabilize a given* setpoint x^*, i.e., $\mathbb{X}^* = \{x^*\}$. On the other hand, in this chapter we are also interested in the stabilization of a possibly unbounded set $\mathbb{X}^*$, such as, e.g., in consensus and synchronization problems. In this case, existence of α_2 satisfying Assumption 3.7 does not necessarily follow from continuity of V^f and the fact that $V^f(x) = 0$ for all $x \in \mathbb{X}^*$, and hence needs to be assumed. Note, however, that this assumption is in general not very restrictive. Furthermore, we remark that the existence of $\alpha_2 \in \mathcal{K}_\infty$ as required by Assumption 3.7 implies that the terminal region $\mathbb{X}^f$ as defined in Assumption 3.6 satisfies the first condition of Assumption 3.5 with $\delta^f = \alpha_2^{-1}(a)$.

3.2 Distributed MPC schemes for general cooperative control problems

In this section, we present two distributed MPC schemes for the solution of cooperative control problems. Both schemes use a special optimization sequence of the systems and are based on the method presented in (Richards and How, 2007), where this idea was used in the context of (robust) distributed stabilization of linear systems including coupling constraints but no couplings in the cost functions (and hence also decentralized terminal regions could be used). The first proposed scheme in Section 3.2.1 is applicable for general cost functions (satisfying Assumptions 3.3–3.7), while for the second one (see Section 3.2.2) the cost functions in addition need to have a certain separable structure, which leads to simplifications and a decrease in the necessary communication between the systems. In Section 3.2.3, we then discuss and compare various properties of the two schemes.

3.2.1 Distributed MPC with general cost functions

The first proposed distributed MPC scheme is specified as follows.

Algorithm 3.4 (Distributed MPC with general cost functions).

0. *Initialization: Set $t = 0$, and for all systems $i \in \mathbb{I}_{[1,P]}$, find[2] a feasible input sequence $\hat{u}_i(0)$ with corresponding state sequence $\hat{x}_i(0)$, such that the constraints (3.4) and (3.6) are satisfied. Each system transmits $\hat{x}_i(0)$ to its neighbors. Go to Step 2.*

1. *At time $t \in \mathbb{I}_{\geq 1}$, each system i computes the candidate input sequence*

$$\hat{u}_i(t) := \left\{ u_i^0(1|t-1), \ldots, u_i^0(N-1|t-1), \kappa_i^f(x_i^0(N|t-1), \tilde{x}_{-i}(N|t-1)) \right\} \quad (3.7)$$

by taking the remaining part of the previous optimal input sequence $u_i^0(t-1)$ and adding the local auxiliary control law κ_i^f, and denotes the corresponding candidate state sequence by $\hat{x}_i(t)$. Each system sends $\hat{x}_i(t)$ to all of its neighbors.

2. *After having received $\hat{x}_j(t)$ from all of its neighbors $j \in \mathcal{N}_i$, each system i initializes $\tilde{x}_{-i}(t)$ with*

$$\tilde{x}_{-i}(t) = \begin{bmatrix} \hat{x}_{i_1}(t) \\ \vdots \\ \hat{x}_{i_{d_i}}(t) \end{bmatrix} =: \hat{x}_{-i}(t) \quad (3.8)$$

---

[2]We refer to Remark 3.5 stated below for a further discussion of how this can be done.

and computes $J_{i,old} := J_{i,N}(x_i(t), \tilde{x}_{-i}(t), \hat{u}_i(t))$ and $V^f_{i,old} := V^f_i(\hat{x}_i(N|t), \tilde{x}_{-i}(N|t))$.

3. *For $i = 1$ to P, system i*

   a) *solves Problem 3.1 and denotes its minimizer by $\boldsymbol{u}^0_{i,test}(t)$, the corresponding state sequence by $\boldsymbol{x}^0_{i,test}(t)$, the optimal value of the cost function by $J_{i,test} := J_{i,N}(x_i(t), \tilde{x}_{-i}(t), \boldsymbol{u}^0_{i,test}(t))$, and $V^f_{i,test} := V^f_i(x^0_i(N|t), \tilde{x}_{-i}(N|t))$.*

   b) *sends $\boldsymbol{x}^0_{i,test}(t)$ to all of its neighbors $j \in \mathcal{N}_i$, which then compute $J_{j,test} := J_{j,N}(x_j(t), \tilde{x}_{-j,test}(t), \bar{u}_j(t))$ and $V^f_{j,test} := V^f_j(\bar{x}_j(N|t), \tilde{x}_{-j,test}(N|t))$. Herein, $\tilde{x}_{-j,test}(t) := \tilde{x}_{-j}(t)$, but with $\hat{x}_i(t)$ replaced by $\boldsymbol{x}^0_{i,test}(t)$. Furthermore, $\bar{u}_j(t) = \boldsymbol{u}^0_j(t)$ and $\bar{x}_j(t) = \boldsymbol{x}^0_j(t)$ if $j < i$, i.e., for all neighbors which already have solved Problem 3.1 at time t; otherwise, $\bar{u}_j(t) = \hat{u}_j(t)$ and $\bar{x}_j(t) = \hat{x}_j(t)$, i.e., for all neighbors which have not yet solved Problem 3.1 at time t.*

   c) *receives $J_{j,test}$, $J_{j,old}$, $V^f_{j,test}$ and $V^f_{j,old}$ from all of its neighbors and checks whether*

   $$J_{i,test} - J_{i,old} + \sum_{j \in \mathcal{N}_i} (J_{j,test} - J_{j,old}) \leq 0, \qquad (3.9)$$

   $$e_i(t) := V^f_{i,test} - V^f_{i,old} + \sum_{j \in \mathcal{N}_i} (V^f_{j,test} - V^f_{j,old}) \leq 0. \qquad (3.10)$$

   If (3.9) and (3.10) hold, set $\boldsymbol{u}^0_i(t) := \boldsymbol{u}^0_{i,test}(t)$, $\boldsymbol{x}^0_i(t) := \boldsymbol{x}^0_{i,test}(t)$, $J_{i,old} := J_{i,test}$, $V^f_{i,old} := V^f_{i,test}$ and flag $:= 1$; otherwise, set $\boldsymbol{u}^0_i(t) := \hat{u}_i(t)$, $\boldsymbol{x}^0_i(t) := \hat{x}_i(t)$ and flag $:= 0$.

   d) *sends flag to all of its neighbors $j \in \mathcal{N}_i$. If flag $= 1$, the neighbors set $\tilde{x}_{-j}(t) := \tilde{x}_{-j,test}(t)$, $J_{j,old} := J_{j,test}$, and $F_{j,old} := F_{j,test}$.*

4. *Each system applies $u_i(t) := u^0_i(0|t)$.*

5. *Set $t := t + 1$ and go to Step 1.*

In Steps 3b) and 3c) of Algorithm 3.4, it is checked whether the input sequence $\boldsymbol{u}^0_{i,test}$, obtained from solving Problem 3.1, leads to an overall decrease in the sum of the optimal value functions. Inequality (3.9) can be interpreted as whether the "benefit" (cost decrease) $J_{i,old} - J_{i,test}$, gained from minimizing the local objective function $J_{i,N}$, is greater than the possible "damage" (cost increase) $\Sigma_{j\in\mathcal{N}_i}(J_{j,test} - J_{j,old})$ that is done within the objective functions of system i's neighbors. We note that a similar idea was also used in a different setting by Maestre et al. (2011). Note that the calculation of $J_{i,old}$ and $J_{j,test}$ in Step 2 and 3b), respectively, is a pure evaluation of the respective cost function, i.e., no optimization problem has to be solved. Hence each system i has to solve Problem 3.1 only once per sampling instant, namely in Step 3a). Furthermore, inequality (3.10) will be used to guarantee in a distributed fashion that the centralized terminal constraint (3.6) is satisfied. Namely, similar to what was discussed above for inequality (3.9), with inequality (3.10) it is checked whether the sum of the terminal cost functions when using $\boldsymbol{u}^0_{i,test}$ is less or equal compared to the case when using the feasible sequence $\hat{u}_i$, which together with the special choice of the terminal region $\mathbb{X}^f$ in Assumption 3.6 can be used to show that the centralized terminal constraint (3.6) is satisfied (see the proof of Theorem 3.8 below for more details).

Remark 3.5. *In Step 0 of Algorithm 3.4, i.e., the initialization phase, we assume that feasible input and corresponding state sequences can be found for each system, such that the constraints (3.4) and the centralized terminal constraint (3.6) are satisfied. One possibility for this would be to solve the corresponding centralized MPC problem, i.e., minimize $\sum_{i=1}^{P} J_{i,N}$ over $\boldsymbol{u}(0) := \{u(0|0), \ldots, u(N-1|0)\}$ subject to the constraints (3.4) for all $i \in \mathbb{I}_{[1,P]}$ and (3.6); this problem could either be solved in a centralized way, or by a suitable distributed optimization algorithm.*

Remark 3.6. *Note that at each time $t \in \mathbb{I}_{\geq 0}$, the feasible state sequences $\hat{\boldsymbol{x}}_j(t)$ with which $\tilde{\boldsymbol{x}}_{-i}(t)$ is initialized in (3.8), are successively replaced by $\boldsymbol{x}_j^0(t)$ within Step 3 of Algorithm 3.4. Hence after all systems have performed Step 3, $\tilde{\boldsymbol{x}}_{-i}(t)$ is given by*

$$\tilde{\boldsymbol{x}}_{-i}(t) = \begin{bmatrix} \boldsymbol{x}_{i_1}^0(t) \\ \vdots \\ \boldsymbol{x}_{i_{d_i}}^0(t) \end{bmatrix} =: \boldsymbol{x}_{-i}^0(t) \tag{3.11}$$

for all $i \in \mathbb{I}_{[1,P]}$. But this means that in Step 1, instead of sending the whole feasible state sequence $\hat{\boldsymbol{x}}_i(t)$ to all neighbors, it would be sufficient to just transmit the last state $\hat{x}_i(N|t) = f(x_i^0(N|t-1), \kappa_i^f(x_i^0(N|t-1), \tilde{x}_{-i}(N|t-1)))$ of this sequence. For $k \in \mathbb{I}_{[0,N-1]}$, we have $\hat{x}_i(k|t) = x_i^0(k|t-1)$, and these states are already known to system i's neighbors as discussed above.

In the following, we examine the properties of the overall closed-loop system

$$x(t+1) = f(x(t), u^0(0|t)), \qquad x(0) = x_0, \tag{3.12}$$

resulting from application of Algorithm 3.4, where $u^0(0|t) := [u_1^0(0|t)^T \ \ldots \ u_P^0(0|t)^T]^T$. To this end, we make the following assumption on the initial feasible input and state sequences provided in Step 0 of Algorithm 3.4.

Assumption 3.8. *For all $x_0 \in \mathbb{X}^f$, the initial input sequences $\hat{\boldsymbol{u}}_i(0)$ and the corresponding initial state sequences $\hat{\boldsymbol{x}}_i(0)$ are such that $\sum_{i=1}^{P} J_{i,N}(x_{i0}, \hat{\boldsymbol{x}}_{-i}(0), \hat{\boldsymbol{u}}_i(0)) \leq \alpha_3(|x_0|_{\mathbb{X}^*})$ for some $\alpha_3 \in \mathcal{K}_\infty$.*

Remark 3.7. *Assumption 3.8 will be needed to establish asymptotic stability of the set $\mathbb{X}^*$ instead of only asymptotic convergence of the closed-loop system (3.12) to $\mathbb{X}^*$. Note that Assumption 3.8 is satisfied if for all $x_0 \in \mathbb{X}^f$, the initial input sequences $\hat{\boldsymbol{u}}_i(0)$ are defined by just using N times the local auxiliary control law κ_i^f for all $i \in \mathbb{I}_{[1,P]}$. Namely, in this case, we obtain*

$$\sum_{i=1}^{P} J_{i,N}(x_{i0}, \hat{\boldsymbol{x}}_{-i}(0), \hat{\boldsymbol{u}}_i(0)) \leq \sum_{i=1}^{P} V_i^f(x_{i0}, x_{-i0}) = V^f(x_0) \leq \alpha_2(|x_0|_{\mathbb{X}^*}), \tag{3.13}$$

where the first inequality follows from N times applying property iv) of Assumption 3.5, and the second inequality follows from Assumption 3.7. Alternatively, Assumption 3.8 is also satisfied if the initial input sequences $\hat{\boldsymbol{u}}_i(0)$ are determined by solving the corresponding centralized MPC problem as described in Remark 3.5. Namely, as the cost of the optimal solution is less or equal than that of the (feasible) one where the local controllers κ_i^f are applied, again (3.13), and hence also Assumption 3.8 (with $\alpha_3 = \alpha_2$), is satisfied. Finally, we remark that Assumption 3.8 was missing in the references (Müller and Allgöwer, 2014a; Müller et al., 2011, 2012b), on which the results of this chapter are based as stated above.

Now denote by $\mathbb{X}_N$ the set of all states $x_0 \in \mathbb{X}$ such that Step 0 of Algorithm 3.4 is feasible. We can then state the following result.

Theorem 3.8. *Suppose that $x_0 \in \mathbb{X}_N$ and that Assumptions 3.1–3.8 are satisfied. Then, Problem 3.1 in Step 3a) of Algorithm 3.4 is feasible for all $t \in \mathbb{I}_{\geq 0}$, for all systems $i \in \mathbb{I}_{[1,P]}$. Furthermore, for the overall closed-loop system (3.12) resulting from application of Algorithm 3.4, it holds that (i) for all $t \in \mathbb{I}_{\geq 0}$, the pointwise-in-time local state and input constraints (3.2) for all $i \in \mathbb{I}_{[1,P]}$ and coupling constraints (3.3) for all $r \in \mathbb{I}_{[1,R]}$ are satisfied, and (ii) the set $\mathbb{X}^*$ is asymptotically stable with region of attraction $\mathbb{X}_N$.*

Proof. As usual in MPC, we first prove recursive feasibility of Algorithm 3.4, before establishing asymptotic stability of the set $\mathbb{X}^*$.

Part 1: The proof of recursive feasibility of Algorithm 3.4 mainly follows the arguments of Richards and How (2007), except for establishing satisfaction of the (centralized) terminal constraint (3.6) in a distributed fashion. First, note that as discussed in Remark 3.6, the feasible state sequences $\hat{\boldsymbol{x}}_j(t)$ with which $\tilde{\boldsymbol{x}}_{-i}(t)$ is initialized in (3.8), are successively replaced by $\boldsymbol{x}_j^0(t)$, so that at the end of the loop in Step 3 we have $\tilde{\boldsymbol{x}}_{-i}(t) = \boldsymbol{x}_{-i}^0(t)$ for all $t \in \mathbb{I}_{\geq 0}$ and all $i \in \mathbb{I}_{[1,P]}$. Thus, for all $t \in \mathbb{I}_{\geq 0}$, when computing the candidate input sequence $\hat{\boldsymbol{u}}_i(t)$ (3.7) in Step 1, each system uses $\tilde{\boldsymbol{x}}_{-i}(N|t-1) = x_{-i}^0(N|t-1)$.

Now assume that for some $t \in \mathbb{I}_{\geq 0}$, the input and corresponding state sequences $\boldsymbol{u}_i^0(t)$ and $\boldsymbol{x}_i^0(t)$, respectively, satisfy the constraints (3.4) (with $\tilde{\boldsymbol{x}}_{-i}(t) = \boldsymbol{x}_{-i}^0(t)$) for all $i \in \mathbb{I}_{[1,P]}$ and the centralized terminal constraint (3.6), i.e., $x^0(N|t) := [x_1^0(N|t)^T \ \ldots \ x_P^0(N|t)^T]^T \in \mathbb{X}^f$. Consider the system with index $i = 1$ at the following time instant[3] $t + 1$. By standard MPC argumentation, the candidate input sequence $\hat{\boldsymbol{u}}_1(t + 1)$ as defined by (3.7) and the corresponding state sequence $\hat{\boldsymbol{x}}_1(t + 1)$ satisfy the local state and input constraints (3.4c). Furthermore, the terminal constraint (3.6) is also satisfied, i.e., $\hat{x}(N|t + 1) := [\hat{x}_1(N|t + 1)^T \ \ldots \ \hat{x}_P(N|t + 1)^T]^T \in \mathbb{X}^f$, as

$$V^f(\hat{x}(N|t + 1)) = V^f\big(f(x^0(N|t), \kappa^f(x^0(N|t)))\big)$$

$$\overset{\text{Ass. 3.5 } iv)}{\leq} V^f(x^0(N|t)) - \sum_{i=1}^{P} \ell_i(x_i^0(N|t), x_{-i}^0(N|t), \kappa_i^f(x_i^0(N|t), x_{-i}^0(N|t)))$$

$$\leq V^f(x^0(N|t)) \leq a, \tag{3.14}$$

where the last inequality follows as $x^0(N|t) \in \mathbb{X}^f$ by assumption. Finally, consider the coupling constraints (3.4d). As for $k = 0, \ldots, N - 1$, both $\hat{x}_1(k|t + 1) = x_1^0(k + 1|t)$ and $\tilde{x}_{-1}(k|t + 1) = x_{-1}^0(k + 1|t)$, the coupling constraints are fulfilled due to the assumption that they were fulfilled at time t for the sequences $\boldsymbol{x}_i^0(t)$. For $k = N$, the coupling constraints (3.4d) are also satisfied according to condition $iii)$ of Assumption 3.5, as $\hat{x}(N|t + 1) \in \mathbb{X}^f$ as established above. Hence the candidate input sequence $\hat{\boldsymbol{u}}_1$ (3.7) is a feasible solution to Problem 3.1 for system 1 at time $t + 1$, and consequently also the state sequence $\boldsymbol{x}_{1,test}^0(t + 1)$, obtained in Step 3a), satisfies the constraints (3.4). Furthermore, according to Step 3c), $\boldsymbol{x}_{1,test}^0(t + 1)$ is only assigned to $\boldsymbol{x}_1^0(t + 1)$ if (3.10) is satisfied, i.e., if the sum of the terminal cost functions when using $\boldsymbol{x}_{1,test}^0(t + 1)$ is less

---

[3]The following considerations are also valid at time $t = 0$, where the candidate input and state sequences $\hat{\boldsymbol{u}}_i(0)$ and $\hat{\boldsymbol{x}}_i(0)$, respectively, are determined in the initialization Step 0 of Algorithm 3.4 and not in Step 1.

or equal than when the feasible state sequence $\hat{x}_1(t+1)$ is used. But then, defining $x' := [x_1^0(N|t+1)^T \ \hat{x}_2(N|t+1)^T \ \dots \ \hat{x}_P(N|t+1)^T]^T$, we obtain

$$V^f(x') \stackrel{(3.10)}{=} V^f(\hat{x}(N|t+1)) + \min\{0, e_1(t+1)\} \stackrel{(3.14)}{\leq} a,$$

which by Assumption 3.6 means that the terminal constraint (3.6) is satisfied, i.e., $x' \in \mathbb{X}^f$.

Assume now that at time $t+1$, systems 1 to i_0 with $i_0 \in \mathbb{I}_{[1,P-1]}$ have already performed Step 3 of Algorithm 3.4, with Problem 3.1 in Step 3a) being feasible and $[x_1^0(N|t+1)^T, \dots, x_{i_0}^0(N|t+1)^T, \hat{x}_{i_0+1}(N|t+1)^T, \dots, \hat{x}_P(N|t+1)^T]^T \in \mathbb{X}^f$. Then, there also exists a feasible solution to Problem 3.1 at time $t+1$ for the system i_0+1. This can be proven as follows. Satisfaction of the state and input constraints (3.4c) for the candidate input sequence $\hat{u}_{i_0+1}(t+1)$ (3.7) and the corresponding candidate state sequence $\hat{x}_{i_0+1}(t+1)$ is established as above for system 1. Furthermore, the candidate state sequence $\hat{x}_{i_0+1}(t+1)$ is exactly the sequence which the previously optimizing neighboring systems $j \in \mathcal{N}_{i_0+1} \cap \mathbb{I}_{[1,i_0]}$ assumed for system i_0+1, i.e., $\hat{x}_{i_0+1}(t+1)$ was used within $\tilde{x}_j(t+1)$, which means that the coupling constraints (3.4d) are satisfied. Hence $\hat{u}_{i_0+1}(t+1)$ is a feasible solution to Problem 3.1 at time $t+1$ for the system i_0+1. Consequently, analogous to what was established above for system 1, with $x_{i_0+1}^0(t+1)$ as defined in Step 3c) the constraints (3.4) are satisfied and furthermore

$$V^f(x'') \stackrel{(3.10)}{=} V^f(\hat{x}(N|t+1)) + \sum_{i=1}^{i_0+1} \min\{0, e_i(t+1)\} \stackrel{(3.14)}{\leq} a, \tag{3.15}$$

where $x'' := [x_1^0(N|t+1)^T, \dots, x_{i_0+1}^0(N|t+1)^T, \hat{x}_{i_0+2}(N|t+1)^T, \dots, \hat{x}_P(N|t+1)^T]^T$. Hence again by Assumption 3.6, also the terminal constraint (3.6) is satisfied, i.e., $x'' \in \mathbb{X}^f$. With this, by induction over i_0, feasibility of Problem 3.1 for all systems $i \in \mathbb{I}_{[1,P]}$ at time $t+1$ can be established, with $x^0(N|t+1)$ satisfying the terminal constraint (3.6). Feasibility for all times then follows by induction over t and the assumption that $x_0 \in \mathbb{X}_N$, i.e., the initialization Step 0 in Algorithm 3.4 is feasible.

Given the above, satisfaction of the pointwise-in-time local state and input constraints (3.2) and coupling constraints (3.3) for the overall closed-loop system (3.12) immediately follows from the definition of the control inputs $u_i(t) := u_i^0(0|t)$ in Step 4 of Algorithm 3.4.

Part 2: We now proceed with proving that the set $\mathbb{X}^*$ is asymptotically stable for the overall closed-loop system (3.12). To this end, consider the functions

$$V_i(t) := J_{i,N}(x_i(t), \boldsymbol{x}_{-i}^0(t), \boldsymbol{u}_i^0(t))$$

and use $V(t) := \sum_{i=1}^P V_i(t)$ as a Lyapunov function candidate. According to Step 3) of Algorithm 3.4, it holds that

$$V(t) = \sum_{i \notin \{\mathcal{N}_P \cup \{P\}\}} V_i(t) + \sum_{i \in \mathcal{N}_P} V_i(t) + V_P(t)$$

$$\leq \sum_{i \notin \{\mathcal{N}_P \cup \{P\}\}} V_i(t) + \sum_{i \in \mathcal{N}_P} J_{i,N}\left(x_i(t), \begin{bmatrix} \boldsymbol{x}_{i_1}^0(t) \\ \vdots \\ \boldsymbol{x}_{i_{d_i-1}}^0(t) \\ \hat{\boldsymbol{x}}_P(t) \end{bmatrix}, \boldsymbol{u}_i^0(t)\right) + J_{P,N}(x_P(t), \boldsymbol{x}_{-P}^0(t), \hat{\boldsymbol{u}}_P(t)),$$

$$\tag{3.16}$$

where the inequality in (3.16) holds due to the definition of $\boldsymbol{x}_P^0(t)$ in Step 3c) of Algorithm 3.4. Namely, the newly calculated locally optimal state sequence $\boldsymbol{x}_{P,test}^0(t)$ is only assigned to $\boldsymbol{x}_P^0(t)$ if (3.9) holds, which establishes the inequality in (3.16). On the other hand, if (3.9) is not satisfied, then $\hat{\boldsymbol{x}}_P(t)$ is assigned to $\boldsymbol{x}_P^0(t)$, and thus (3.16) holds with equality. Using the same argument as in (3.16) recursively from $i = P$ down to 1, one obtains that

$$V(t) \leq \sum_{i=1}^{P} J_{i,N}(x_i(t), \hat{\boldsymbol{x}}_{-i}(t), \hat{\boldsymbol{u}}_i(t)). \tag{3.17}$$

Thus, along the closed-loop system (3.12) with $x_0 \in \mathbb{X}_N$, for all $t \in \mathbb{I}_{\geq 0}$ we have

$$V(t+1) - V(t) \overset{(3.17)}{\leq} \sum_{i=1}^{P} J_{i,N}(x_i(t+1), \hat{\boldsymbol{x}}_{-i}(t+1), \hat{\boldsymbol{u}}_i(t+1)) - V(t)$$

$$= \sum_{i=1}^{P} \Big(- \ell_i(x_i^0(0|t), x_{-i}^0(0|t), u_i^0(0|t)) + \ell_i(x_i^0(N|t), x_{-i}^0(N|t), \kappa_i^f(x_i^0(N|t), x_{-i}^0(N|t))) \Big)$$

$$+ V^f(f(x^0(N|t), \kappa^f(x^0(N|t)))) - V^f(x^0(N|t))$$

$$\overset{\text{Ass. 3.5 } iv)}{\leq} - \sum_{i=1}^{P} \ell_i\big(x_i^0(0|t), x_{-i}^0(0|t), u_i^0(0|t)\big),$$

$$\overset{\text{Ass. 3.4}}{\leq} -\alpha_1(|x^0(0|t)|_{\mathbb{X}^*}) = -\alpha_1(|x(t)|_{\mathbb{X}^*}), \tag{3.18}$$

where for the second inequality we used the fact that $x^0(N|t) \in \mathbb{X}^f$, as established above. From here, using Lyapunov arguments, it follows as in standard MPC stability proofs (see, e.g., Theorem 2.22 in Rawlings and Mayne (2009)) that the overall closed-loop system (3.12) asymptotically converges to the set $\mathbb{X}^*$, for all $x_0 \in \mathbb{X}_N$. Furthermore, using again similar arguments as in standard MPC stability results (Rawlings and Mayne, 2009, Theorem 2.22), one can show that the set $\mathbb{X}^*$ is stable. Namely, due to Assumptions 3.4 and 3.7, we obtain $V(t) \geq \sum_{i=1}^{P} \ell_i(x_i(t), x_{-i}(t), u_i(t)) \geq \alpha_1(|x(t)|_{\mathbb{X}^*})$ for all $t \in \mathbb{I}_{\geq 0}$ and all $x_0 \in \mathbb{X}_N$. Moreover, from (3.17) and Assumption 3.8 it follows that $V(0) \leq \sum_{i=1}^{P} J_{i,N}(x_{i0}, \hat{\boldsymbol{x}}_{-i}(0), \hat{\boldsymbol{u}}_i(0)) \leq \alpha_3(|x_0|_{\mathbb{X}^*})$ for all $x_0 \in \mathbb{X}^f$, which together with (3.18) implies that $V(t) \leq \alpha_3(|x_0|_{\mathbb{X}^*})$ for all $t \in \mathbb{I}_{\geq 0}$ and all $x_0 \in \mathbb{X}^f$. As $\mathbb{X}^f \subseteq \mathbb{X}_N$, this means that for all $x_0 \in \mathbb{X}^f$ we have

$$|x(t)|_{\mathbb{X}^*} \leq \alpha_1^{-1}(V(t)) \leq \alpha_1^{-1}(\alpha_3(|x_0|_{\mathbb{X}^*})). \tag{3.19}$$

By Assumption 3.5, there exists a $\delta^f > 0$ such that $|x_0|_{\mathbb{X}^*} \leq \delta^f$ and $x_0 \in \mathbb{X}$ implies $x_0 \in \mathbb{X}^f$. But then, from (3.19) it follows that for each $\varepsilon > 0$ there exists a $\delta > 0$, namely $\delta := \min\{\delta^f, \alpha_3^{-1}(\alpha_1(\varepsilon))\}$, such that $|x_0|_{\mathbb{X}^*} \leq \delta$ and $x_0 \in \mathbb{X}$ implies that $|x(t)|_{\mathbb{X}^*} \leq \varepsilon$ for all $t \in \mathbb{I}_{\geq 0}$, i.e., the set $\mathbb{X}^*$ is stable. Summarizing the above, we have established that $\mathbb{X}^*$ is asymptotically stable for the overall closed-loop system (3.12) resulting from application of Algorithm 3.4 with region of attraction $\mathbb{X}_N$, which concludes the proof of Theorem 3.8. $\square$

As pointed out earlier, in Algorithm 3.4 the centralized terminal constraint (3.6) could not be directly incorporated into Problem 3.1, but its satisfaction was ensured in a distributed way via (3.10). This might result in an unnecessary use of the candidate input sequences $\hat{\boldsymbol{u}}_i$ instead of the newly calculated optimal sequences $\boldsymbol{u}_{i,test}^0$ in Step 3c) of Algorithm 3.4. We will show in Section 3.2.2 how this limitation can be removed if the cost functions exhibit a certain separable structure.

3.2.2 Distributed MPC with separable cost functions

In this section, we show how Algorithm 3.4 can be simplified if the stage and terminal cost functions have a certain separable structure. Namely, in this case an overall decrease in the sum of the optimal value functions does not have to be ensured via (3.9); furthermore, as already pointed out above, the centralized terminal constraint (3.6) can be separated and consequently directly included into the optimization problems 3.1, and hence does not have to be ensured via (3.10). To this end, we make the following assumption on the stage and terminal cost functions.

Assumption 3.9. *For all $i \in \mathbb{I}_{[1,P]}$, the stage and terminal cost functions ℓ_i and V_i^f, respectively, are of the form*

$$\ell_i(x_i, x_{-i}, u_i) := \ell_{ii}(x_i, u_i) + \sum_{j \in \mathcal{N}_i} \ell_{ij}(x_i, x_j),$$
$$V_i^f(x_i, x_{-i}) := V_{ii}^f(x_i) + \sum_{j \in \mathcal{N}_i} V_{ij}^f(x_i, x_j), \qquad (3.20)$$

for some $\ell_{ii} : \mathbb{X}_i \times \mathbb{U}_i \to \mathbb{R}$, $\ell_{ij} : \mathbb{X}_i \times \mathbb{X}_j \to \mathbb{R}$, $V_i^f : \mathbb{X}_i \to \mathbb{R}$, and $V_{ij}^f : \mathbb{X}_i \times \mathbb{X}_j \to \mathbb{R}$.

Assumption 3.9 means that the stage and terminal cost functions of each system are separated into a part ℓ_{ii} (respectively, V_{ii}^f) depending on its own state and input, and parts ℓ_{ij} (respectively, V_{ij}^f) depending on its own and *one* of its neighbors' states. In order to make use of this separable structure, we define the following modified optimization problem for each system i:

Problem 3.9.

$$\operatorname*{minimize}_{\boldsymbol{u}_i(t)} \bar{J}_{i,N}(x_i(t), \tilde{\boldsymbol{x}}_{-i}(t), \boldsymbol{u}_i(t))$$

subject to the constraints (3.4) and the additional terminal constraint

$$x_i(N|t) \in \mathbb{X}_i^f(t), \qquad (3.21)$$

where

$$\bar{J}_{i,N}(x_i(t), \tilde{\boldsymbol{x}}_{-i}(t), \boldsymbol{u}_i(t)) := J_{i,N}(x_i(t), \tilde{\boldsymbol{x}}_{-i}(t), \boldsymbol{u}_i(t)) + \sum_{k=0}^{N-1} \sum_{j \in \mathcal{N}_i} \ell_{ji}(\tilde{x}_j(k|t), x_i(k|t))$$
$$+ \sum_{j \in \mathcal{N}_i} V_{ji}^f(\tilde{x}_j(N|t), x_i(N|t)). \qquad (3.22)$$

The (time-varying) terminal regions $\mathbb{X}_i^f(k)$ appearing in (3.21) will be specified below in Algorithm 3.10. Furthermore, the cost function $\bar{J}_{i,N}$ as defined by (3.22) is comprised of system i's own cost function, $J_{i,N}$ (defined as in Problem 3.1), and those parts of the cost functions of its neighbors which involve its own state x_i. This requires that system i knows the functions ℓ_{ji} and V_{ji}^f, for all $j \in \mathcal{N}_i$, which is assumed in the following. The modified distributed MPC algorithm can now be specified as follows.

Algorithm 3.10 (Distributed MPC with separable cost functions).

0. *Same as Step 0 of Algorithm 3.4.*

1. *Same as Step 1 of Algorithm 3.4.*

2. *After having received $\hat{\boldsymbol{x}}_j(t)$ from all of its neighbors $j \in \mathcal{N}_i$, each system i initializes $\tilde{\boldsymbol{x}}_{-i}(t)$ with $\tilde{\boldsymbol{x}}_{-i}(t) = \hat{\boldsymbol{x}}_{-i}(t)$ as given in (3.8).*

3. *For $i = 1$ to P, system i*

   a) defines the terminal region

   $$\mathbb{X}_i^f(t) := \Big\{ y \in \mathbb{X}_i : V_i^f(y, \tilde{x}_{-i}(N|t)) + \sum_{j \in \mathcal{N}_i} V_{ji}^f(\tilde{x}_j(N|t), y)$$
   $$\leq V_i^f(\hat{x}_i(N|t), \tilde{x}_{-i}(N|t)) + \sum_{j \in \mathcal{N}_i} V_{ji}^f(\tilde{x}_j(N|t), \hat{x}_i(N|t)) \Big\}, \qquad (3.23)$$

   b) solves Problem 3.9 with $\mathbb{X}_i^f(t)$ given by (3.23), and denotes its minimizer by $\boldsymbol{u}_i^0(t)$ and the corresponding state sequence by $\boldsymbol{x}_i^0(t)$,

   c) sends $\boldsymbol{x}_i^0(t)$ to its neighbors $j \in \mathcal{N}_i$, which update $\tilde{\boldsymbol{x}}_{-j}(t)$ by replacing $\hat{\boldsymbol{x}}_i(t)$ with $\boldsymbol{x}_i^0(t)$.

4. *Each system applies $u_i(t) := u_i^0(0|t)$.*

5. *Set $t := t + 1$ and go to Step 1.*

We now show that Algorithm 3.10 enjoys the same properties as Algorithm 3.4.

Theorem 3.11. *Suppose that $x_0 \in \mathbb{X}_N$ and that Assumptions 3.1–3.9 are satisfied. Then, Problem 3.9 in Step 3b) of Algorithm 3.10 is feasible for all $t \in \mathbb{I}_{\geq 0}$, for all systems $i \in \mathbb{I}_{[1,P]}$. Furthermore, for the overall closed-loop system (3.12) resulting from application of Algorithm 3.10, it holds that (i) for all $t \in \mathbb{I}_{\geq 0}$, the pointwise-in-time local state and input constraints (3.2) for all $i \in \mathbb{I}_{[1,P]}$ and coupling constraints (3.3) for all $r \in \mathbb{I}_{[1,R]}$ are satisfied, and (ii) the set $\mathbb{X}^*$ is asymptotically stable with region of attraction $\mathbb{X}_N$.*

Proof. Recursive feasibility of Problem 3.9 can be shown as in the proof of Theorem 3.8, with the exception that the terminal constraint is now directly included into Problem 3.9. Clearly, the candidate state sequence $\hat{\boldsymbol{x}}_i(t)$ satisfies the terminal constraint (3.21) and (3.23), and hence Problem 3.9 is recursively feasible for all systems $i \in \mathbb{I}_{[1,P]}$. Furthermore, note that the time-varying distributed terminal regions $\mathbb{X}_i^f(t)$ are defined such that the centralized terminal constraint (3.6) is satisfied, i.e., $x^0(N|t) \in \mathbb{X}^f$. Namely, due to the separable structure (3.20) of V_i^f, the value $e_i(t)$ as defined in (3.10) is given by

$$e_i(t) = V_i^f(x_i^0(N|t), \tilde{x}_{-i}(N|t)) - V_i^f(\hat{x}_i(N|t), \tilde{x}_{-i}(N|t))$$
$$+ \sum_{j \in \mathcal{N}_i} \big(V_{ji}^f(\tilde{x}_j(N|t), x_i^0(N|t)) - V_{ji}^f(\tilde{x}_j(N|t), \hat{x}_i(N|t)) \big), \qquad (3.24)$$

and $e_i(t) \leq 0$ according to the definition of $\mathbb{X}_i^f(t)$ in (3.23) and the fact that $x_i^0(N|t) \in \mathbb{X}_i^f(t)$. But then, similar to what was shown in the proof of Theorem 3.8 (compare (3.15) with $i_0 = P - 1$), for all $t \in \mathbb{I}_{\geq 0}$ we obtain

$$V^f(x^0(N|t)) = V^f(\hat{x}(N|t)) + \sum_{i=1}^{P} e_i(t) \leq V^f(\hat{x}(N|t)) \leq a, \tag{3.25}$$

and hence $x^0(N|t) \in \mathbb{X}^f$ according to Assumption 3.6.

Satisfaction of the pointwise-in-time local state and input constraints (3.2) and coupling constraints (3.3) for the overall closed-loop system (3.12) again immediately follows from the definition of the control inputs $u_i(t) := u_i^0(0|t)$ in Step 4 of Algorithm 3.10.

In order to prove that the set $\mathbb{X}^*$ is asymptotically stable for the overall closed-loop system (3.12), consider the following. For any system i and any neighbor $h \in \mathcal{N}_i$, define

$$\ell_i^h(x_i, x_{-i}, u_i) := \ell_{ii}(x_i, u_i) + \sum_{j \in \mathcal{N}_i, j \neq h} \ell_{ij}(x_i, x_j) = \ell_i(x_i, x_{-i}, u_i) - \ell_{ih}(x_i, x_h),$$

$$V_i^{fh}(x_i, x_{-i}) := V_{ii}^f(x_i) + \sum_{j \in \mathcal{N}_i, j \neq h} V_{ij}^f(x_i, x_j) = V_i^f(x_i, x_{-i}) - V_{ih}^f(x_i, x_h),$$

$$J_{i,N}^h(x_i(t), \tilde{\boldsymbol{x}}_{-i}(t), \boldsymbol{u}_i(t)) := \sum_{k=0}^{N-1} \ell_i^h\big(x_i(k|t), \tilde{x}_{-i}(k|t), u_i(k|t)\big) + V_i^{fh}\big(x_i(N|t), \tilde{x}_{-i}(N|t)\big).$$
$$\tag{3.26}$$

Now consider again the functions

$$V_i(t) := J_{i,N}(x_i(t), \boldsymbol{x}_{-i}^0(t), \boldsymbol{u}_i^0(t))$$

and use $V(t) := \sum_{i=1}^{P} V_i(t)$ as a Lyapunov function candidate. Note that while the modified cost function $\bar{J}_{i,N}$ is used within Problem 3.9, we still use $J_{i,N}$ in the definition of V_i. We obtain

$$V(t) = \sum_{i \notin \{\mathcal{N}_P \cup \{P\}\}} V_i(t) + \sum_{i \in \mathcal{N}_P} V_i(t) + V_P(t)$$

$$\stackrel{(3.26)}{=} \sum_{i \notin \{\mathcal{N}_P \cup \{P\}\}} V_i(t) + \sum_{i \in \mathcal{N}_P} J_{i,N}^P(x_i(t), \boldsymbol{x}_{-i}^0(t), \boldsymbol{u}_i^0(t)) + \sum_{k=0}^{N-1} \sum_{i \in \mathcal{N}_P} \ell_{iP}(x_i^0(k|t), x_P^0(k|t))$$

$$+ \sum_{i \in \mathcal{N}_P} V_{iP}^f(x_i^0(N|t), x_P^0(N|t)) + J_{P,N}(x_P(k), \boldsymbol{x}_{-P}^0(t), \boldsymbol{u}_P^0(t))$$

$$\stackrel{(3.22)}{=} \sum_{i \notin \{\mathcal{N}_P \cup \{P\}\}} V_i(t) + \sum_{i \in \mathcal{N}_P} J_{i,N}^P(x_i(t), \boldsymbol{x}_{-i}^0(t), \boldsymbol{u}_i^0(t)) + \bar{J}_{P,N}(x_P(k), \boldsymbol{x}_{-P}^0(t), \boldsymbol{u}_P^0(t)).$$
$$\tag{3.27}$$

But as in Step 3b) of Algorithm 3.10, system P solves Problem 3.9 with objective function $\bar{J}_{P,N}$, due to optimality of $\boldsymbol{u}_P^0$ we obtain

$$\bar{J}_{P,N}(x_P(t), \boldsymbol{x}_{-P}^0(t), \boldsymbol{u}_P^0(t)) \leq \bar{J}_{P,N}(x_P(t), \boldsymbol{x}_{-P}^0(t), \hat{\boldsymbol{u}}_P(t))$$

$$\stackrel{(3.22)}{=} J_{P,N}(x_P(t), \boldsymbol{x}_{-P}^0(t), \hat{\boldsymbol{u}}_P(t)) + \sum_{k=0}^{N-1} \sum_{i \in \mathcal{N}_P} \ell_{iP}(x_i^0(k|t), \hat{x}_P(k|t)) + \sum_{i \in \mathcal{N}_i} V_{iP}^f(x_i^0(N|t), \hat{x}_P(N|t)).$$

Plugging this into (3.27) and using again (3.26) yields

$$V(t) \leq \sum_{i \notin \{\mathcal{N}_P \cup \{P\}\}} V_i(t) + \sum_{i \in \mathcal{N}_P} J_{i,N}(x_i(t), \begin{bmatrix} \boldsymbol{x}_{i_1}^0(t) \\ \vdots \\ \boldsymbol{x}_{i_{d_{i-1}}}^0(t) \\ \hat{\boldsymbol{x}}_P(t) \end{bmatrix}, \boldsymbol{u}_i^0(t)) + J_{P,N}(x_P(t), \boldsymbol{x}_{-P}^0(t), \hat{\boldsymbol{u}}_P(t)).$$

(3.28)

Using the above argument as in (3.28) recursively from $i = P$ down to 1, one again obtains that (3.17) holds, and the rest of the proof then follows along the lines of the proof of Theorem 3.8. $\qquad\square$

3.2.3 Discussion

In this section, we discuss and compare various properties of Algorithms 3.4 and 3.10. We start with some remarks on the communication requirements and scalability of the algorithms, before discussing a relaxation in the way in which the centralized terminal constraint (3.6) is ensured, in order to reduce possible conservatism.

First, note that Algorithms 3.4 and 3.10 are non-iterative, i.e., each system solves its optimization problem (Problems 3.1 and 3.9, respectively) only *once* at each time t, namely in Step 3a) of Algorithm 3.4 and Step 3b) of Algorithm 3.10, respectively. The advantage of such an approach is that less communication and computational power is needed than if the systems had to exchange predicted sequences and solve their respective optimization problems multiple times at each time t. On the other hand, the price we have to pay is that we cannot expect to achieve a performance which is as good as if the centralized problem for the overall system was solved iteratively.

The communication requirement for Algorithm 3.4 is such that at each time t, each system sends data to each of its neighbors four times. Namely, once in Step 1, where each system sends the feasible state sequence $\hat{\boldsymbol{x}}_i$ to all neighbors (or, as discussed in Remark 3.6, just its last state $\hat{x}_i(N|t)$), then in Steps 3b) and 3d) in order to determine $\boldsymbol{u}_i^0(t)$, and finally in Step 3c) when one of its neighbors determines $\boldsymbol{u}_j^0(t)$. On the other hand, in Algorithm 3.10, the communication load is significantly reduced, as each system has to send data to its neighbors only twice at each time t, namely in Steps 1 and 3c). This is due to the separable structure of the functions ℓ_i and V_i^f and the special choice of the cost function $\bar{J}_{i,N}$ which is minimized within Problem 3.9. Namely, as shown in the proof of Theorem 3.11, this allows us to guarantee a decrease in V and satisfaction of the centralized terminal constraint (3.6) without the need to exchange $J_{j,test}$, $J_{j,old}$, $V_{j,test}^f$ and $V_{j,old}^f$ as in Algorithm 3.4, and one also does not have to possibly use the candidate (fallback) solution $\hat{\boldsymbol{u}}_i$ as was the case in Algorithm 3.4.

For clarity of presentation, in Step 3 of both Algorithms 3.4 and 3.10, we required the systems to solve their optimization problem in a specified, sequential order. This special sequential structure enabled us to ensure recursive feasibility as well as asymptotic stability of the set $\mathbb{X}^*$, as shown in Theorems 3.8 and 3.11, respectively. However, if the systems have to execute Step 3 in a sequential order during one sampling interval, this implies that the respective distributed MPC schemes are not very well scalable with the total number of systems P. Nevertheless, it is straightforward to show that Theorems 3.8 and 3.11 are still valid if in Algorithms 3.4 and 3.10, respectively, two systems i and j for which $\mathcal{N}_i \cap \mathcal{N}_j = \emptyset$ perform Step 3 in parallel. For Algorithm 3.10, this can be further relaxed to allowing non-neighboring systems to execute Step 3 in parallel[4]. This means that the scalability of the algorithm depends on the underlying graph of the network. For example, if the underlying graph is a tree (i.e., does not contain cycles) or, more general, a bipartite graph (Godsil and Royle, 2001), then the systems can be divided into two disjoint independent sets (i.e., sets which contain only non-neighboring nodes, see Godsil and Royle (2001)), independent of the total number of systems P. This is also known as a proper 2-coloring of the graph (Godsil and Royle, 2001). Hence, in case of Algorithm 3.10, only two optimization problems have to be solved sequentially at each time t independent of the number of systems P, i.e., the algorithm is scalable. For general graphs, d optimization problems have to be solved sequentially in Algorithm 3.10 during each time t, where d is the minimum number such that the graph can be properly d-colored, also known as the *chromatic number* of the graph (Godsil and Royle, 2001). Furthermore, we note that while determining the chromatic number of a graph is in general an NP-hard problem, (polynomial time) approximate graph coloring algorithms exist with a certain guaranteed upper bound on the maximum number of colors (independent sets), see, e.g., Halldórsson (1993). In summary, scalability of Algorithms 3.4 and 3.10 depends on the interconnection structure of the systems, and in particular on the number of independent sets into which the corresponding graph can be partitioned.

Finally, note that ensuring satisfaction of the centralized terminal constraint (3.6) as done in Algorithm 3.4 (via (3.10)) and Algorithm 3.10 (via (3.21) and (3.23)) can be conservative, as in fact this requires that the sum of the terminal cost functions, V^f, decreases at each time instant. Namely, by inspecting the proofs of Theorems 3.8 and 3.11, one can see that in both cases, for all $t \in \mathbb{I}_{\geq 0}$ we obtain

$$V^f(x^0(N|t+1)) = V^f(\hat{x}(N|t+1)) + \sum_{i=1}^{P} \min\{0, e_i(t+1)\}$$
$$\overset{(3.14)}{\leq} V^f((x^0(N|t))) - \sum_{i=1}^{P} \ell_i(x_i^0(N|t), x_{-i}^0(N|t), \kappa_i^f(x_i^0(N|t), x_{-i}^0(N|t)))$$
$$+ \sum_{i=1}^{P} \min\{0, e_i(t+1)\}, \tag{3.29}$$

where for Algorithm 3.4 the first equality follows from (3.15) (with $i_0 = P-1$), and from (3.25) for Algorithm 3.10. On the other hand, for satisfaction of the centralized terminal constraint (3.6), according to Assumption 3.6 it would be sufficient that $V^f(x^0(N|t)) \leq a$ for all $t \in \mathbb{I}_{\geq 0}$. For Algorithm 3.4, a way to reduce the possible conservatism is as follows. For all $i \in \mathbb{I}_{[1,P]}$, let $\bar{e}_i(t) := e_i(t) + \hat{e}_i(t)$ with $e_i(t)$ as defined in (3.10) and

$$\hat{e}_i(t+1) := \begin{cases} \bar{e}_i(t) - \ell_i(x_i^0(N|t), x_{-i}^0(N|t), \kappa_i^f(x_i^0(N|t), x_{-i}^0(N|t))) & \text{if } \bar{e}_i(t) \leq 0 \\ \hat{e}_i(t) - \ell_i(x_i^0(N|t), x_{-i}^0(N|t), \kappa_i^f(x_i^0(N|t), x_{-i}^0(N|t))) & \text{else} \end{cases} \tag{3.30}$$

---

[4]Note that the latter situation is not valid for Algorithm 3.4, but we need the stricter requirement that only systems for which $\mathcal{N}_i \cap \mathcal{N}_j = \emptyset$ can perform Step 3 in parallel. This is needed in Steps 3b-c), where we have to ensure that two systems i_1 and i_2, which are both neighbors of a system j, do not request the computation of $J_{j,test}$ and $V_{j,test}^f$ from system j at the same time.

for all $t \in \mathbb{I}_{\geq 0}$, and $\hat{e}_i(0) \leq 0$ such that $\sum_{i=1}^{P} \hat{e}_i(0) = V^f(\hat{x}(N|0)) - a$. Then, require-
ment (3.10) in Step 3c) of Algorithm 3.4 is replaced by instead checking whether $\bar{e}_i(t) \leq 0$.
This means that the "buffer" $\hat{e}_i$ allows the sum of the terminal cost functions to possibly
increase, at most by the amount it has decreased at earlier time steps. With this modification, the proof of Theorem 3.8 still works in the same way, i.e., its results still hold. For
Algorithm 3.10, a similar relaxation can be used. Namely, instead of using $\mathbb{X}_i^f$ as given
by (3.23), one can define the modified terminal region

$$\overline{\mathbb{X}}_i^f(t) := \Big\{ y \in \mathbb{X}_i : V_i^f(y, \tilde{x}_{-i}(N|t)) + \sum_{j \in \mathcal{N}_i} V_{ji}^f(\tilde{x}_j(N|t), y) + \tilde{e}_i(t)$$

$$\leq V_i^f(\hat{x}_i(N|t), \tilde{x}_{-i}(N|t)) + \sum_{j \in \mathcal{N}_i} V_{ji}^f(\tilde{x}_j(N|t), \hat{x}_i(N|t)) \Big\}. \tag{3.31}$$

Here, for $t \in \mathbb{I}_{\geq 0}$, $\tilde{e}_i(t+1) := e_i(t) + \tilde{e}_i(t) - \ell_i(x_i^0(N|t), x_{-i}^0(N|t), \kappa_i^f(x_i^0(N|t), x_{-i}^0(N|t)))$
with $e_i(t)$ given by (3.24), and $\tilde{e}_i(0) \leq 0$ such that $\sum_{i=1}^{P} \tilde{e}_i(0) = V^f(\hat{x}(N|0)) - a$. Again,
with this modification, the proof of Theorem 3.11 still works in the same way, i.e., its
results still hold. To summarize, the two relaxations described above result in an improved
performance in Algorithms 3.4 and 3.10, as they result in a less restrictive inequality (3.10)
and a larger terminal region $\overline{\mathbb{X}}_i^f(t)$, respectively.

3.3 Application to consensus and synchronization

The two distributed MPC schemes as presented in Sections 3.2.1 and 3.2.2 are formulated in
a way such that they are suitable for a variety of cooperative control problems. As already
mentioned above, one special case is $\mathbb{X}^* = \{x^*\}$ for some given setpoint x^*. This means
that the cooperative control task consists of asymptotically stabilizing the overall system at
the (a priori) given setpoint x^*, while satisfying coupling constraints and minimizing some
possibly coupled cost function. As discussed in Section 1.2.1, this is the control objective
which is considered in most of the available distributed MPC schemes in the literature.
In (Müller et al., 2012b, Section 4), it was shown in more detail how Algorithms 3.4
and 3.10 can be used for this special cooperative control task, and that in fact decoupled
terminal regions can be obtained in this case.

In this Section, we show how Algorithms 3.4 and 3.10 can be used for more general
cooperative control tasks than setpoint stabilization. In particular, we exemplarily study
the control objective of consensus and synchronization, i.e., we want to synchronize the
states of P identical systems. As noted in Section 1.2.1, this is an important cooperative
control problem appearing in many different applications. For conciseness, we focus on
identical linear systems, i.e., (3.1) reduces to

$$x_i(t+1) = Ax_i(t) + Bu_i(t), \qquad x_i(0) = x_{i0}, \tag{3.32}$$

with $x_i(t) \in \mathbb{X}_i \subseteq \mathbb{R}^n$ and $u_i(t) \in \mathbb{U}_i \subseteq \mathbb{R}^m$ for all $t \in \mathbb{I}_{\geq 0}$ and all $i \in \mathbb{I}_{[1,P]}$, $A \in$
$\mathbb{R}^{n \times n}$, and $B \in \mathbb{R}^{n \times m}$. For extensions of the presented results to nonlinear systems, the
interested reader is referred to (Müller et al., 2012b, Section 5.2). The control objective
of synchronizing the systems (3.32) can now be translated into the stabilization of the set
$\mathbb{X}^* = \{x \in \mathbb{R}^{Pn} : x_1 = x_2 = \cdots = x_P\}$. In the following, we show how the stage cost

functions ℓ_i and the required terminal ingredients, i.e., the terminal cost functions V_i^f and the local auxiliary control laws κ_i^f, can be defined such that the required assumptions (in particular Assumption 3.5) are satisfied. To this end, we make use of the incidence matrix $E(\mathcal{G}) \in \mathbb{R}^{P \times Q}$ of the graph $\mathcal{G}$ specifying the interconnection topology of the systems, which we assume to be connected; here, $Q := |\mathcal{E}|$ is the number of edges of the graph $\mathcal{G}$. In order to define $E(\mathcal{G})$, we assign an arbitrary orientation to each edge, i.e. such that each edge has a head and a tail. The rows of $E(\mathcal{G})$ are now indexed by the set of nodes, and the columns by the set of edges, and the iq-entry of $E(\mathcal{G})$ takes the value $+1$ if node i is the head of edge q, -1 if it is the tail, and 0 otherwise, see, e.g., (Godsil and Royle, 2001, p.167) and Zelazo and Mesbahi (2011). Define $\xi_q := x_i - x_j$ as the state difference of the systems i and j corresponding to the q-th column of $E(\mathcal{G})$ indexed by the edge (v_i, v_j). Let $\xi := [\xi_1^T \ \ldots \ \xi_Q^T]^T$, and note that $\xi = (E(\mathcal{G})^T \otimes I_n)x$ and $\mathbb{X}^*$ can be expressed as $\mathbb{X}^* = \{x \in \mathbb{R}^{Pn} : \xi = 0\}$. Now consider the local auxiliary control law candidates

$$\kappa_i^f(x_i, x_{-i}) = \sum_{q \in \mathcal{N}_i^h} K_q \xi_q - \sum_{q \in \mathcal{N}_i^t} K_q \xi_q, \qquad (3.33)$$

where $\mathcal{N}_i^h$ denotes the set of edges with node i being the head, $\mathcal{N}_i^t$ the set of edges with node i being the tail, and $K_q \in \mathbb{R}^{m \times n}$ are the local auxiliary controller gains to be determined. Note that κ_i^f as defined in (3.33) only depends on system i's own state, x_i, and neighboring states x_{-i}, as required. From (3.33), it follows that $\kappa^f := [(\kappa_1^f)^T \ \ldots \ (\kappa_P^f)^T]^T$ can be expressed as

$$\kappa^f(x) = (E(\mathcal{G}) \otimes I_m)K\xi = (E(\mathcal{G}) \otimes I_m)K(E(\mathcal{G})^T \otimes I_n)x, \qquad (3.34)$$

where $K = \text{diag}(K_1, \ldots, K_Q)$. With this, the overall system in closed-loop with the local auxiliary control law can be written as

$$\begin{aligned} x(t+1) &= (I_P \otimes A)x(t) + (I_P \otimes B)\kappa^f(x(t)) \\ &= (I_P \otimes A)x(t) + (I_P \otimes B)(E(\mathcal{G}) \otimes I_m)K(E(\mathcal{G})^T \otimes I_n)x(t) \\ &= \Big((I_P \otimes A) + (I_P \otimes B)(E(\mathcal{G}) \otimes I_m)K(E(\mathcal{G})^T \otimes I_n)\Big)x(t). \end{aligned} \qquad (3.35)$$

Furthermore, by noting that

$$(E(\mathcal{G})^T \otimes I_n)(I_P \otimes A) = E(\mathcal{G})^T I_P \otimes I_n A = I_Q E(\mathcal{G})^T \otimes A I_n = (I_Q \otimes A)(E(\mathcal{G})^T \otimes I_n), \qquad (3.36)$$

we obtain from (3.35) that

$$\begin{aligned} \xi(t+1) &= (E(\mathcal{G})^T \otimes I_n)x(t+1) \\ &= (E(\mathcal{G})^T \otimes I_n)\Big((I_P \otimes A) + (I_P \otimes B)(E(\mathcal{G}) \otimes I_m)K(E(\mathcal{G})^T \otimes I_n)\Big)x(t) \\ &= \Big((I_Q \otimes A) + (E(\mathcal{G})^T \otimes I_n)(I_P \otimes B)(E(\mathcal{G}) \otimes I_m)K\Big)\xi(t) =: \Psi\xi(t). \end{aligned} \qquad (3.37)$$

Now for each system $i \in \mathbb{I}_{[1,P]}$, consider the stage and terminal cost functions given by

$$\ell_i(x_i, x_{-i}, u_i) = u_i^T S_i u_i + \sum_{q \in \mathcal{N}_i^t \cup \mathcal{N}_i^h} \xi_q^T T_q \xi_q, \qquad (3.38)$$

$$V_i^f(x_i, x_{-i}) = \sum_{q \in \mathcal{N}_i^t \cup \mathcal{N}_i^h} \xi_q^T W_q \xi_q, \qquad (3.39)$$

for some positive definite matrices $S_i \in \mathbb{R}^{m \times m}$ and $T_q, W_q \in \mathbb{R}^{n \times n}$. Note that ℓ_i and V_i^f given by (3.38) and (3.39), respectively, satisfy Assumptions 3.3, 3.4, 3.7 and 3.9. In order to be able to apply Theorems 3.8 and 3.11, it remains to determine the local auxiliary controller gains K_q as well as the terminal weighting matrices W_q such that Assumption 3.5 is satisfied. With $W := \mathrm{diag}(W_1, \ldots, W_Q)$, $T := \mathrm{diag}(T_1, \ldots, T_Q)$, and $S := \mathrm{diag}(S_1, \ldots, S_P)$, we obtain

$$V^f(x) = \sum_{i=1}^{P} V_i^f(x_i, x_{-i}) = \sum_{i=1}^{P} \sum_{q \in \mathcal{N}_i^t \cup \mathcal{N}_i^h} \xi_q^T W_q \xi_q = 2\xi^T W \xi, \tag{3.40}$$

$$\sum_{i=1}^{P} \ell_i(x_i, x_{-i}, \kappa_i^f(x_i, x_{-i})) = \sum_{i=1}^{P} \kappa_i^f(x_i, x_{-i})^T S_i \kappa_i^f(x_i, x_{-i}) + \sum_{i=1}^{P} \sum_{q \in \mathcal{N}_i^t \cup \mathcal{N}_i^h} \xi_q^T T_q \xi_q$$

$$= \kappa^f(x)^T S \kappa^f(x) + 2\xi^T T \xi. \tag{3.41}$$

With this, according to Assumption 3.6, the terminal region $\mathbb{X}^f$ is given by

$$\mathbb{X}^f = \{x \in \mathbb{R}^{Pn} : 2\xi^T W \xi \le a\} \tag{3.42}$$

for some $a > 0$. Furthermore, Assumption 3.5iv) translates into

$$V^f(f(x, \kappa^f(x))) - V^f(x) + \sum_{i=1}^{P} \ell_i(x_i, x_{-i}, \kappa_i^f(x_i, x_{-i}))$$

$$\stackrel{(3.34),(3.37),(3.40),(3.41)}{=} \xi^T \Big(2\Psi^T W \Psi - 2W + K^T (E(\mathcal{G}) \otimes I_m)^T S (E(\mathcal{G}) \otimes I_m) K + 2T \Big) \xi \le 0,$$

for all $x \in \mathbb{X}^f$. Denoting $\tilde{A} := (I_Q \otimes A)$, $\tilde{B} := (E(\mathcal{G})^T \otimes I_n)(I_P \otimes B)$ and $\tilde{K} := (E(\mathcal{G}) \otimes I_m) K$, this results in

$$\xi^T \Big(2(\tilde{A} + \tilde{B}\tilde{K})^T W (\tilde{A} + \tilde{B}\tilde{K}) - 2W + \tilde{K}^T S \tilde{K} + 2T \Big) \xi \le 0 \tag{3.43}$$

for all ξ. Inequality (3.43) can be rewritten as the following equivalent linear matrix inequality (LMI) with $X := W^{-1} \succ 0$ and $Y := \tilde{K} X$ by using standard manipulations such as the Schur complement and left and right multiplying with W^{-1} (see, e.g., Kothare et al. (1996); Maestre et al. (2011)):

$$\begin{bmatrix} 2X & \sqrt{2}(X\tilde{A}^T + Y^T \tilde{B}^T) & \sqrt{2}X T^{1/2} & Y^T S^{1/2} \\ \sqrt{2}(\tilde{A}X + \tilde{B}Y) & X & 0_{Qn \times Qn} & 0_{Qn \times Pm} \\ \sqrt{2}T^{1/2}X & 0_{Qn \times Qn} & I_{Qn} & 0_{Qn \times Pm} \\ S^{1/2}Y & 0_{Pm \times Qn} & 0_{Pm \times Qn} & I_{Pm} \end{bmatrix} \succeq 0. \tag{3.44}$$

Due to the block-diagonal structure of W, also X is block-diagonal and thus Y has the same structure as $\tilde{K}$. Hence $\tilde{K}$ can be recovered from the solution of (3.44) as $\tilde{K} = Y X^{-1} = Y W$, from which K can be obtained by recalling that $\tilde{K}$ was defined as $\tilde{K} := (E(\mathcal{G}) \otimes I_m) K$ and K is block-diagonal, which means that the q-th diagonal block of K can be recovered from the q-th block-column of $\tilde{K}$ due to the specific structure of the incidence matrix $E(\mathcal{G})$.

In case that the graph $\mathcal{G}$ contains cycles, some of the state differences ξ_q are linearly dependent, namely those corresponding to a circle subgraph. To be more precise, the linear dependencies of the state differences ξ_q are given by $e_{null}^T \xi = 0$, where e_{null} is a vector in the nullspace of $E(\mathcal{G}) \otimes I_n$ (compare Godsil and Royle (2001); Zelazo and Mesbahi (2011)). Let us quickly illustrate this property with a simple example. Consider the cycle graph C_3 with three nodes and edges and incidence matrix given by

$$E(C_3) = \begin{bmatrix} 1 & 1 & 0 \\ -1 & 0 & 1 \\ 0 & -1 & -1 \end{bmatrix}, \qquad (3.45)$$

i.e., $\xi_1 = x_1 - x_2$, $\xi_2 = x_1 - x_3$, and $\xi_3 = x_2 - x_3$. Clearly, it holds that $\xi_2 = \xi_1 + \xi_3$, i.e., $e_{null}^T \xi = [I_n \ -I_n \ I_n] \xi = 0$, where $e_{null} = [I_n \ -I_n \ I_n]^T$ spans the nullspace of $E(C_3) \otimes I_n$.

Considering the above, it follows that in order for Assumption 3.5iv) to be satisfied, (3.43) has to hold only for those ξ such that $E_{null}^T \xi = 0$, where E_{null} is a basis for the nullspace of $E(\mathcal{G}) \otimes I_n$. Hence, also the LMI (3.44) can be relaxed as follows. Namely, according to Finsler's Lemma (Finsler (1937), see also Boyd et al. (1994)), (3.43) being satisfied for all ξ such that $E_{null}^T \xi = 0$ is equivalent to the existence of a constant $\rho > 0$ such that

$$2(\tilde{A} + \tilde{B}\tilde{K})^T W(\tilde{A} + \tilde{B}\tilde{K}) - 2W + \tilde{K}^T S\tilde{K} + 2T - \rho E_{null} E_{null}^T \preceq 0. \qquad (3.46)$$

However, (3.46) cannot be transformed into an LMI analogously to (3.44), as the additional term $\rho E_{null} E_{null}^T$ in (3.46) would result in an additional term $\rho X E_{null} E_{null}^T X$ in the first diagonal element of (3.44), which is quadratic in X and hence destroys the LMI structure. Nevertheless, we can tighten (3.46) in the following way in order to still obtain an LMI:

$$2(\tilde{A} + \tilde{B}\tilde{K})^T W(\tilde{A} + \tilde{B}\tilde{K}) - 2W + \tilde{K}^T S\tilde{K} + 2T - \rho(W E_{null} E_{null}^T + E_{null} E_{null}^T W) \preceq 0 \qquad (3.47)$$

for some $\rho > 0$. Note that solvability of (3.47) implies solvability of (3.46), while the converse is in general not true. Namely, (3.47) implies that (3.43) holds for all ξ such that $E_{null}^T \xi = 0$, which, as explained above, is equivalent to (3.46) according to Finsler's Lemma; hence solvability of (3.47) implies solvability of (3.46). Now (3.47) can be transformed into an equivalent LMI (for fixed ρ) analogously to above, and we obtain

$$\begin{bmatrix} 2X + \rho(E_{null} E_{null}^T X + X E_{null} E_{null}^T) & \sqrt{2}(X\tilde{A}^T + Y^T \tilde{B}^T) & \sqrt{2} X T^{1/2} & Y^T S^{1/2} \\ \sqrt{2}(\tilde{A}X + \tilde{B}Y) & X & 0_{Qn \times Qn} & 0_{Qn \times Pm} \\ \sqrt{2} T^{1/2} X & 0_{Qn \times Qn} & I_{Qn} & 0_{Qn \times Pm} \\ S^{1/2} Y & 0_{Pm \times Qn} & 0_{Pm \times Qn} & I_{Pm} \end{bmatrix} \succeq 0. \qquad (3.48)$$

Note that although (3.47) (and hence also the equivalent LMI (3.48)) is more restrictive than (3.46), it still offers a considerable relaxation compared to the LMI (3.44), where the linear dependency of the state differences ξ_q due to the cycles in the graph was not considered. Furthermore, note that in the case when the graph contains no cycles, it follows that E_{null} is empty, and hence from (3.48) we recover the LMI (3.44). Again, as explained above, the local auxiliary controller gains K_q and the terminal weighting matrices W_q can be recovered from the solution X and Y of the LMI (3.48). We thus arrive at the following result:

Proposition 3.12. *Suppose that the LMI (3.48) is solvable for some $\rho > 0$, and that the local auxiliary control laws κ_i^f and the terminal cost functions V_i^f are computed according to (3.33) and (3.39), respectively, where K_q and W_q are obtained from the solution of the LMI (3.48). Then Assumption 3.5iv) is satisfied.*

Finally, it remains to specify the value of a, which defines the size of the terminal region in (3.42). Namely, $a > 0$ has to be chosen small enough such that $\mathbb{X}^f \subseteq \mathbb{X}$ and conditions i) and iii) of Assumption 3.5 are satisfied. Hence we assume that the constraint sets $\mathbb{X}_i$, $\mathbb{Z}_i$, $\mathbb{C}_r$, and the functions c_r are such that there exists an $a > 0$ such that this is possible; note that this is a very reasonable assumption. Namely, for coupling constraints which match the control objective of reaching consensus, one can assume that they are satisfied in a neighborhood of the consensus subspace $\mathbb{X}^*$, i.e., that Assumption 3.5iii) is satisfied there. Furthermore, existence of an $a > 0$ such that condition i) of Assumption 3.5 is satisfied is, e.g., guaranteed if $\mathbb{Z} = \mathbb{X} \times \mathbb{U}$ and $0 \in \text{int}(\mathbb{U})$.

Summarizing the above, we have shown how the stage cost functions ℓ_i and the required terminal ingredients, i.e., the terminal cost functions V_i^f and the local auxiliary control laws κ_i^f, can be defined for the control objective of reaching consensus, i.e., $\mathbb{X}^* = \{x \in \mathbb{R}^{Pn} : x_1 = x_2 = \cdots = x_P\}$. Hence, we can apply both Algorithms 3.4 and 3.10 to conclude via Theorems 3.8 and 3.11, respectively, that $\mathbb{X}^*$ is asymptotically stable for the resulting closed-loop system (3.12), i.e., the systems asymptotically reach consensus.

Remark 3.13. *In case that $\mathbb{X}^* = \{x \in \mathbb{R}^{Pn} : x_1 = x_2 = \cdots = x_P\}$ is not contained in[5] $\mathbb{Z}_{\mathbb{X}}$, which is in particular the case if some of the constraint sets $\mathbb{Z}_i$ are bounded, the control objective is rather to stabilize the set $\widetilde{\mathbb{X}}^* := \mathbb{X}^* \cap \mathbb{Z}_{\mathbb{X}}$ (or some set $\widetilde{\mathbb{X}}^* \subseteq \mathbb{X}^* \cap \mathbb{Z}_{\mathbb{X}}$), i.e., the consensus subspace intersected with the state constraint set to be satisfied. In this case, one also has to use a modified terminal region $\widetilde{\mathbb{X}}^f := \mathbb{X}^f \cap \overline{\mathbb{X}}$ for some $\overline{\mathbb{X}} := \overline{\mathbb{X}}_1 \times \cdots \times \overline{\mathbb{X}}_P \subseteq \mathbb{Z}_{\mathbb{X}}$ and $\mathbb{X}^f$ given by (3.42). However, in contrast to $\mathbb{X}^f$, the modified terminal region $\widetilde{\mathbb{X}}^f$ does not satisfy Assumption 3.6. Nevertheless, the results of Theorems 3.8 and 3.11 are still valid in this case if in Problems 3.1 and 3.9, respectively, the additional constraint $x_i(N|t) \in \overline{\mathbb{X}}_i$ is added and furthermore Assumptions 3.5, 3.7 and 3.8 are satisfied with $\mathbb{X}^f$ replaced by $\widetilde{\mathbb{X}}^f$. While the above analysis still ensures that Assumption 3.5iv) is satisfied for the modified terminal region $\widetilde{\mathbb{X}}^f$ with κ_i^f and V_i^f as defined in (3.33) and (3.39), respectively, note that the invariance condition ii) of Assumption 3.5 is not necessarily implied anymore by condition iv). Whether Assumption 3.5ii) is satisfied for the modified terminal region $\widetilde{\mathbb{X}}^f$ with κ_i^f as defined in (3.33) will in general depend on the specific shape of the involved constraint sets $\mathbb{Z}_i$ and $\mathbb{C}_r$, and the system dynamics (3.32).*

3.3.1 Example - synchronization of linear oscillators

To illustrate our results, we consider the problem of synchronizing six two-dimensional identical linear oscillators. The dynamics of each oscillator is given by (3.32) with $x_i = [x_{1i} \ x_{2i}]^T \in \mathbb{R}^2$, $u_i \in \mathbb{R}$, and system matrices $A = \begin{bmatrix} 0.9762 & 0.2169 \\ -0.2169 & 0.9762 \end{bmatrix}$ and $B = \begin{bmatrix} 0 & 1 \end{bmatrix}^T$. The interconnection topology of the oscillators is specified by the graph $\mathcal{G}$, which is depicted in Figure 3.1 together with the state differences ξ_q corresponding to the columns of $E(\mathcal{G})$. The sets $\mathbb{Z}_i$ are given as $\mathbb{Z}_i := \mathbb{R}^2 \times [-1, 1]$, and the prediction horizon is $N = 15$. Figure 3.2

---

[5]Recall that $\mathbb{Z}_{\mathbb{X}}$ denotes the projection of $\mathbb{Z}$ on $\mathbb{X}$.

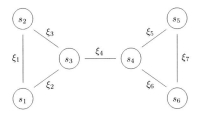

Figure 3.1: Interconnection topology of the linear oscillators in Section 3.3.1.

shows simulation results obtained by applying Algorithm 3.10, with initial conditions $x_1(0) = \begin{bmatrix} 5 & 3 \end{bmatrix}^T$, $x_2(0) = \begin{bmatrix} 3 & 1 \end{bmatrix}^T$, $x_3(0) = \begin{bmatrix} 0 & 4 \end{bmatrix}^T$, $x_4(0) = \begin{bmatrix} -2 & 4 \end{bmatrix}^T$, $x_5(0) = \begin{bmatrix} -3 & -3 \end{bmatrix}^T$, and $x_6(0) = \begin{bmatrix} -2 & -4 \end{bmatrix}^T$. The local auxiliary control laws κ_i^f and the terminal cost functions V_i^f are computed according to (3.33) and (3.39), respectively, where K_q and W_q are obtained from the solution of the LMI (3.48). The input weights S_i for the stage cost functions ℓ_i in (3.38) are chosen as $S_i = 1$ for all $i \in \mathbb{I}_{[1,6]}$, and different weights T_q for the state differences ξ_q are used in Figures 3.2(a) and 3.2(b). In Figure 3.2(a), the least weight is put on the state difference between systems 3 and 4; this results in the fact that the two subnetworks consisting of systems $1, 2, 3$ and $4, 5, 6$, respectively, synchronize faster than the overall network. On the other hand, in Figure 3.2(b), the largest weight is put on the state difference between systems 3 and 4; in this case, the overall network behavior is such that the state sequences of systems 3 and 4 are much closer to each other.

3.4 Summary

In this chapter, we presented two distributed MPC schemes, where the first one can be used for general coupled cost functions, while the second requires a special separable structure which results in various simplifications. We showed that for both schemes, one can establish recursive feasibility as well as asymptotic stability of the target set $\mathbb{X}^*$ for the resulting closed-loop system. Furthermore, we discussed several properties such as the required communication load between the systems and implementation and scalability issues. One of the main features of both schemes is that they can not only be used for setpoint stabilization, but are also suitable for more general cooperative control tasks. This means that the presented results can in particular be seen as a contribution towards the use of MPC for cooperative control problems in networks of interacting systems. As a particular application, we showed how the developed distributed MPC schemes can be used in consensus and synchronization problems, which we also illustrated with a numerical example.

There are various open research questions arising from the results presented in this chapter. For example, it would be interesting to examine the distributed MPC schemes in light of disturbances and uncertainties, or communication imperfections such as packet losses or delays. Concerning the former, a first result was obtained by (Müller et al., 2012c; Schürmann, 2012) for linear systems subject to additive disturbances. For a more detailed discussion on the latter issues as well as further possible future research topics, we refer the reader to Section 6.2 at the end of this thesis.

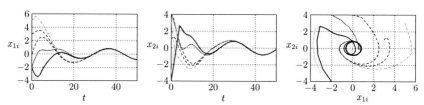

(a) With weights $T_1 = T_2 = T_3 = T_5 = T_6 = T_7 = 2I_2$ and $T_4 = 0.1I_2$; systems $1 - 3$: dashed curves, systems $4 - 6$: solid curves.

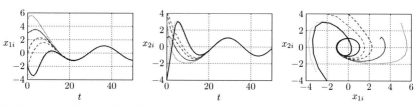

(b) With weights $T_1 = T_2 = T_3 = T_5 = T_6 = T_7 = 0.1I_2$ and $T_4 = 2I_2$; systems 1, 2, 5 and 6 solid curves, systems 3 and 4: dashed curves.

Figure 3.2: Synchronization of linear oscillators via Algorithm 3.10. Left and middle columns: time domain (first and second state components), right column: phase plane

Chapter 4

Economic MPC: Dissipativity, average constraints, and self-tuning terminal cost

In this chapter, we present several contributions to the field of economic MPC. Namely, as motivated and discussed in Sections 1.3 and 2.2, we first thoroughly investigate the role of dissipativity in economic MPC (see Section 4.1), before considering economic MPC including average constraints in Section 4.2. Finally, in Section 4.3 we develop an economic MPC framework with a self-tuning terminal cost.

The results presented in this chapter are based on (Müller and Allgöwer, 2012; Müller et al., 2013a,b,c,d, 2014b,c,d,f).

4.1 Dissipativity in economic MPC

As shown in Section 2.2, dissipativity plays a crucial role in economic MPC, both for classifying optimal steady-state operation and for stability analysis of the closed-loop system. The contribution of this section is to investigate in detail this dissipativity condition; in particular, we examine whether dissipativity is necessary and not only sufficient for optimal steady-state operation (Section 4.1.1), and we establish robustness properties for changing constraint sets (Section 4.1.2). Note that the results presented in this section are important for and find application in settings with terminal constraints (as presented in Section 2.2), but also in settings without such terminal constraints (see, e.g., Grüne (2013)). This is the case as the dissipativity condition is used in both settings and is formulated independently of the specific economic MPC framework.

4.1.1 Necessity of dissipativity for optimal steady-state operation

In Theorem 2.9, it was shown that dissipativity with respect to the supply rate $s(x, u) = \ell(x, u) - \ell(x^*, u^*)$ is a sufficient condition for a system of the form (2.1) to be optimally operated at steady-state. In this section, we examine whether this condition is not only sufficient, but also necessary. This question is important in order to decide whether dissipativity is only a conservative or rather a precise characterization of the property of steady-state optimality. We first show by means of a counterexample that the above dissipativity property is not necessary in the most general case. Nevertheless, under rather mild controllability assumptions on the system, dissipativity with respect to the supply rate $s(x, u) = \ell(x, u) - \ell(x^*, u^*)$ is in fact necessary for optimal steady-state operation. In order to establish these results, we make use of the concept of available storage as introduced in (Willems, 1972, Definition 3), which we slightly adapt to our setting with state and input constraints. To this end, recall the definition of $\mathbb{Z}^0$ in (2.13) and the corresponding projected set $\mathbb{X}^0$.

Definition 4.1. *For a given supply rate* $s : \mathbb{Z}^0 \to \mathbb{R}$ *and each* $x \in \mathbb{X}^0$, *the* available storage S_a *of system* (2.1) *subject to state and input constraints of the form* $(x, u) \in \mathbb{Z}^0$ *is defined as*

$$S_a(x) := \sup_{\substack{T \geq 0 \\ z(0)=x, \ z(t+1)=f(z(t),v(t)) \\ (z(t),v(t))\in\mathbb{Z}^0 \ \forall t\in\mathbb{I}_{\geq 0}}} \sum_{t=0}^{T-1} -s(z(t), v(t)). \tag{4.1}$$

Note that $S_a(x)$ is nonnegative for all $x \in \mathbb{X}^0$, as $T = 0$ is allowed in (4.1) and by convention the empty sum is zero. As the name suggests, the available storage can be interpreted as the maximum stored "energy" which can be extracted from system (2.1), in case that the supply rate s and the storage function λ are interpreted as the energy supplied to and stored by the system, respectively (see (Willems, 1972) for more details). In (Willems, 1972), it was shown that the available storage plays a crucial role for establishing dissipativity of a system.

Theorem 4.2 (Willems (1972), Theorem 1)**.** *System* (2.1) *is dissipative on* $\mathbb{Z}^0$ *with respect to the supply rate* s *if and only if* $S_a(x) < \infty$ *for all* $x \in \mathbb{X}^0$. *Moreover,* S_a *is a storage function according to Definition 2.7, Equation* (2.11).

Remark 4.3. *The original proof in (Willems, 1972) was done for continuous-time systems without state and input constraints; however, it can straightforwardly be adapted to our setting of discrete-time systems with state and input constraints. Furthermore, as already mentioned in Remark 2.8, the definition of dissipativity in (Willems, 1972) is such that the storage function* λ *is required to be nonnegative, while we require that it is bounded from below on* $\mathbb{X}^0$. *Again, the proof in (Willems, 1972) can straightforwardly be adapted to this modified setting. Finally, Theorem 4.2 can also be adapted to the setting where the storage function* λ *is required to be bounded on* $\mathbb{X}^0$. *Namely, system* (2.1) *is dissipative on* $\mathbb{Z}^0$ *with respect to the supply rate* s *and with a storage function* λ *which is bounded on* $\mathbb{X}^0$ *if and only if* S_a *is bounded on* $\mathbb{X}^0$, *i.e.,* $S_a(x) \leq c < \infty$ *for all* $x \in \mathbb{X}^0$ *and some* $c \geq 0$.

A counterexample

Recall the definition of optimal steady-state operation of system (2.1) from Definition 2.6. The definition of the available storage in (4.1) with supply rate $s(x, u) = \ell(x, u) - \ell(x^*, u^*)$ implies that for each $x \in \mathbb{X}^0$ and for each state/input sequence pair $(\boldsymbol{z}, \boldsymbol{v})$ satisfying $z(0) = x$, $z(t + 1) = f(z(t), v(t))$ and $(z(t), v(t)) \in \mathbb{Z}^0$ for all $t \in \mathbb{I}_{\geq 0}$, we have

$$\limsup_{T \to \infty} \sum_{t=0}^{T-1} -(\ell(z(t), v(t)) - \ell(x^*, u^*)) \leq S_a(x).$$

This inequality can equivalently be stated as

$$\liminf_{T \to \infty} \sum_{t=0}^{T-1} (\ell(z(t), v(t)) - \ell(x^*, u^*)) \geq -S_a(x). \tag{4.2}$$

Now if $S_a(x) < \infty$ and hence $\lim_{T\to\infty} S_a(x)/T = 0$ for all $x \in \mathbb{X}^0$, it follows from (4.2) that (2.9) is satisfied. Hence, dissipativity on $\mathbb{Z}^0$ with respect to the supply rate $s(x,u) = \ell(x,u) - \ell(x^*,u^*)$ is sufficient for system (2.1) to be optimally operated at steady-state, as stated in Theorem 2.9. However, the converse is not necessarily true, which will be illustrated with the following example.

Example 4.4. *Consider the scalar system*

$$x(t+1) = \left(\frac{1}{2} - u(t)^2\right)x(t) \qquad x(0) = x_0 \tag{4.3}$$

with input and state constraint sets $\mathbb{U} := [-1/\sqrt{2}, 1/\sqrt{2}]$ *and* $\mathbb{X} := [-x_{\max}, x_{\max}]$ *for some* $0 < x_{max} < 1$, *respectively, and* $\mathbb{Z} := \mathbb{X} \times \mathbb{U}$. *From (4.3) it immediately follows that the set* $\mathbb{Z}^0$ *defined in (2.13) is given by* $\mathbb{Z}^0 = \mathbb{Z}$, *and hence* $\mathbb{X}^0 = \mathbb{X}$. *The cost function* ℓ *is given by* $\ell(x,u) = u^2 + \log(2)/\log(|x|)$ *for all* $x \in \mathbb{X}$ *with* $x \neq 0$, *continuously extended at* $x = 0$ *with* $\ell(0,u) = u^2$. *The set of feasible state/input equilibrium pairs is computed according to (2.3) as* $S = \{(x,u) : x = 0, u \in \mathbb{U}\}$, *and the optimal state/input equilibrium pair is* $(x^*, u^*) = (0,0)$. *Denote by* $\tilde{x}$ *the state sequence resulting from applying the constant input* $\tilde{u}(t) = 0$ *for all* $t \in \mathbb{I}_{\geq 0}$ *to system (4.3). Solving the system equation, we obtain that* $\tilde{x}(t) = x_0/2^t$. *It is easy to see that each feasible state sequence* $\boldsymbol{x}$ *of system (4.3) satisfies* $0 \leq |x(t)| \leq |\tilde{x}(t)|$ *for all* $t \in \mathbb{I}_{\geq 0}$, *and hence each feasible solution to system (4.3) asymptotically converges to* $x^* = 0$. *But this means that (2.9) is satisfied for each feasible solution, and hence system (4.3) is optimally operated at steady-state. On the other hand, for each* $0 < x_0 < x_{\max}$, *from the definition of the available storage in (4.1) we obtain*

$$S_a(x_0) \geq \limsup_{T\to\infty} \sum_{t=0}^{T-1} -\big(\ell(\tilde{x}(t), \tilde{u}(t)) - \ell(x^*,u^*)\big) = \limsup_{T\to\infty} \sum_{t=0}^{T-1} -\frac{\log(2)}{\log(2^{-t}x_0)}$$

$$= \limsup_{T\to\infty} \sum_{t=0}^{T-1} -\frac{\log(2)}{-t\log(2) + \log(x_0)} = \limsup_{T\to\infty} \sum_{t=0}^{T-1} \frac{1}{t - \log(x_0)/\log(2)} = \infty,$$

where the last equality follows from the fact that the harmonic series is divergent. Hence by Theorem 4.2 the system is not dissipative on $\mathbb{Z}^0$ *with respect to the supply rate* $s(x,u) = \ell(x,u) - \ell(x^*,u^*)$.

Necessity of dissipativity under a controllability/reachability condition

One can also construct similar examples where the lack of dissipativity despite optimal operation at steady-state does not result from a cost function which is "very steep" at the optimal steady-state as in Example 4.4, but is due to a slow convergence rate (non-exponential) to the optimal steady-state (see Müller et al. (2013a)). All in common to these examples is that the lack of dissipativity results from a lack of controllability of the considered system. In the following, we show that dissipativity with respect to the supply rate $s(x,u) = \ell(x,u) - \ell(x^*,u^*)$ is in fact necessary for optimal steady-state operation under a certain controllability/reachability condition. To this end, we need the following definitions. For a given $M \in \mathbb{I}_{\geq 1}$ and each $y \in \mathbb{X}$, denote by $\mathcal{C}_M \subseteq \mathbb{X}$ the set of states which can be steered to the optimal steady-state x^* in M steps in a feasible way, i.e.,

$$\mathcal{C}_M := \{x \in \mathbb{X} : \exists \boldsymbol{v} : \mathbb{I}_{[0,M-1]} \to \mathbb{U} \text{ s.t. } z(0) = x, \ z(t+1) = f(z(t), v(t)),$$
$$(z(t), v(t)) \in \mathbb{Z} \ \forall t \in \mathbb{I}_{[0,M-1]}, \ z(M) = x^*\}. \tag{4.4}$$

Next, let $\mathcal{R}_M$ be the set of states which can be reached from the optimal steady-state x^* in M steps in a feasible way, i.e.,

$$\mathcal{R}_M := \{x \in \mathbb{X} : \exists \boldsymbol{v} : \mathbb{I}_{[0,M-1]} \to \mathbb{U} \text{ s.t. } z(0) = x^*, \; z(t+1) = f(z(t), v(t)),$$
$$(z(t), v(t)) \in \mathbb{Z} \; \forall t \in \mathbb{I}_{[0,M-1]}, \; z(M) = x\}. \quad (4.5)$$

Note that $\mathcal{C}_M \cap \mathcal{R}_M \neq \emptyset$, as by definition x^* is contained in both $\mathcal{C}_M$ and $\mathcal{R}_M$. Now define the set $\mathcal{Z}_M$ as the set of state/input pairs which are part of a feasible state/input sequence pair $(\boldsymbol{z}, \boldsymbol{v})$ which is such that $z(t)$ is in the intersection of $\mathcal{C}_M$ and $\mathcal{R}_M$ for all $t \in \mathbb{I}_{\geq 0}$:

$$\mathcal{Z}_M := \Big\{ (x,u) \in \mathbb{Z} : \exists \boldsymbol{v} : \mathbb{I}_{\geq 0} \to \mathbb{U} \text{ s.t. } (z(0), v(0)) = (x,u), \; z(t+1) = f(z(t), v(t)),$$
$$(z(t), v(t)) \in \mathbb{Z}, \; z(t) \in \mathcal{C}_M \cap \mathcal{R}_M \; \forall t \in \mathbb{I}_{\geq 0}, \Big\} \subseteq \mathbb{Z}^0. \quad (4.6)$$

Note that $\mathcal{Z}_M$ is compact for each $M \in \mathbb{I}_{\geq 1}$ if Assumptions 2.1 and 2.2 are satisfied. Namely, from closedness of $\mathbb{Z}$ and continuity of f, it follows that both $\mathcal{C}_M$ and $\mathcal{R}_M$ are closed. Furthermore, by compactness of $\mathbb{U}$, $\mathcal{R}_M$ is compact; hence also $\mathcal{Z}_M$ is compact. Finally, denote the projection of $\mathcal{Z}_M$ on $\mathbb{X}$ by $\mathcal{X}_M$, i.e.,

$$\mathcal{X}_M := \{x \in \mathbb{X} : \exists u \in \mathbb{U} \text{ s.t. } (x,u) \in \mathcal{Z}_M\}. \quad (4.7)$$

Note that $\mathcal{X}_M \subseteq \mathcal{C}_M \cap \mathcal{R}_M \subseteq \mathcal{X}_{2M}$, where the first inequality follows from the definition of $\mathcal{Z}_M$ and the second follows from the fact that for each $y \in \mathcal{C}_M \cap \mathcal{R}_M$, there exist a feasible state sequence with $x(0) = x^*$, $x(M) = y$, and $x(2M) = x^*$, which implies that $y \in \mathcal{X}_{2M}$. We can now state the following result concerning necessity of dissipativity for optimal steady-state operation.

Theorem 4.5. *Suppose that system* (2.1) *is optimally operated at steady-state and ℓ is bounded on[1] $\mathcal{Z}_M$ for all $M \in \mathbb{I}_{\geq 1}$. Then, system* (2.1) *is dissipative on $\mathcal{Z}_M$ with respect to the supply rate $s(x,u) = \ell(x,u) - \ell(x^*, u^*)$, for all $M \in \mathbb{I}_{\geq 1}$.*

Proof. Fix an arbitrary $M \in \mathbb{I}_{\geq 1}$. For simplicity and without loss of generality, we assume in the following that $\ell(x^*, u^*) = 0$. Assume for contradiction that the system is optimally operated at steady-state, but it is not dissipative on $\mathcal{Z}_M$ with respect to the supply rate $s(x,u) = \ell(x,u) - \ell(x^*, u^*)$. Applying Theorem 4.2 with $\mathbb{Z}^0$ and $\mathbb{X}^0$ replaced by $\mathcal{Z}_M$ and $\mathcal{X}_M$, respectively, it follows that this is equivalent to the fact that the available storage (see Definition 4.1 with $\mathbb{Z}^0$ replaced by $\mathcal{Z}_M$) is infinite for some $y \in \mathcal{X}_M$, and hence

$$\inf_{\substack{T \geq 0 \\ z(0) = y \\ z(t+1) = f(z(t), v(t)) \\ (z(t), v(t)) \in \mathcal{Z}_M \; \forall t \in \mathbb{I}_{\geq 0}}} \sum_{t=0}^{T-1} \ell(z(t), v(t)) = -\infty.$$

This means that for each $r \geq 0$, there exists a state/input sequence pair $(\boldsymbol{x}_r, \boldsymbol{u}_r)$ together with a time instant $T_r \in \mathbb{I}_{\geq 0}$, such that $x_r(0) = y$, $(x_r(t), u_r(t)) \in \mathcal{Z}_M$ for all $t \in \mathbb{I}_{\geq 0}$ and

$$\sum_{t=0}^{T_r-1} \ell(x_r(t), u_r(t)) \leq -r. \quad (4.8)$$

---

[1] Note that ℓ is bounded on $\mathcal{Z}_M$ for all $M \in \mathbb{I}_{\geq 1}$ if, e.g., Assumptions 2.1 and 2.2 are satisfied and ℓ is continuous, as $\mathcal{Z}_M$ is compact in this case as discussed above.

Now consider some $r \geq \underline{r} := 1 + 2M \sup_{(x,u) \in \mathcal{Z}_{2M}} \ell(x,u)$. Note that $\underline{r} < \infty$ as ℓ is bounded on $\mathcal{Z}_{2M}$ by assumption. By definition of $\mathcal{Z}_M$, we have $x_r(T_r) \in \mathcal{R}_M \cap \mathcal{C}_M$. Hence there exists a state/input sequence pair $(\boldsymbol{x}', \boldsymbol{u}')$ satisfying $x'(0) = x_r(T_r)$, $x'(M) = x^*$, $x'(2M) = x_r(0) = y$ and $(x'(t), u'(t)) \in \mathcal{Z}$ for all $t \in \mathbb{I}_{[0,2M]}$. As both $x'(0) = x_r(T_r) \in \mathcal{X}_M$ and $x'(2M) = y \in \mathcal{X}_M$, it follows that $(x'(t), u'(t)) \in \mathcal{Z}_{2M}$ for all $t \in \mathbb{I}_{[0,2M]}$.

Now define the following input sequence:

$$\hat{u}(k(T_r + 2M) + i) = \begin{cases} u_r(i) & k \in \mathbb{I}_{\geq 0}, i \in \mathbb{I}_{[0,T_r-1]} \\ u'(i) & k \in \mathbb{I}_{\geq 0}, i \in \mathbb{I}_{[T_r,T_r+2M-1]} \end{cases} \tag{4.9}$$

which results in a cyclic state sequence with $\hat{x}(k(T_r + 2M)) = y$ for all $k \in \mathbb{I}_{\geq 0}$. This state/input sequence pair fulfills $(\hat{x}(t), \hat{u}(t)) \in \mathcal{Z}_{2M} \subseteq \mathbb{Z}^0$ for all $t \in \mathbb{I}_{\geq 0}$ by construction, and furthermore we obtain for all $k \in \mathbb{I}_{\geq 0}$:

$$\sum_{i=0}^{T_r+2M-1} \ell\Big(\hat{x}(k(T_r + 2M) + i), \hat{u}(k(T_r + 2M) + i)\Big) \stackrel{(4.9)}{=} \sum_{i=0}^{T_r-1} \ell(x_r(i), u_r(i)) + \sum_{i=0}^{2M-1} \ell(x'(i), u'(i))$$

$$\stackrel{(4.8)}{\leq} -r + 2M \sup_{(x,u) \in \mathcal{Z}_{2M}} \ell(x,u) \leq -1.$$

But this implies that

$$\liminf_{T \to \infty} \sum_{t=0}^{T-1} \frac{\ell(\hat{x}(t), \hat{u}(t))}{T} \stackrel{(4.9)}{=} \frac{1}{T_r + 2M} \sum_{i=0}^{T_r+2M-1} \ell(\hat{x}(i), \hat{u}(i)) \leq -\frac{1}{T_r + 2M} < 0$$

contradicting (2.9), i.e., optimal steady-state operation. Hence we conclude that the system (2.1) is dissipative on $\mathcal{Z}_M$ with respect to the supply rate $s(x,u) = \ell(x,u) - \ell(x^*, u^*)$. □

Combining Theorems 2.9 and 4.5, we arrive at the following corollary.

Corollary 4.6. *Suppose that $\mathcal{Z}_M = \mathbb{Z}^0$ for some $M \in \mathbb{I}_{\geq 1}$ and ℓ is bounded on $\mathbb{Z}^0$. Then system (2.1) is optimally operated at steady-state if and only if it is dissipative on $\mathbb{Z}^0$ with respect to the supply rate $s(x,u) = \ell(x,u) - \ell(x^*, u^*)$.*

It is easy to show that $\mathcal{Z}_M = \mathbb{Z}^0$ for some $M \in \mathbb{I}_{\geq 1}$ if and only if $\mathcal{X}_M = \mathbb{X}^0$ for some $M \in \mathbb{I}_{\geq 1}$, which means that the system is weakly reversible (Sontag, 1998, Section 4.3) in $\mathbb{X}^0$ in finite time, i.e., each $x' \in \mathbb{X}^0$ can be reached from any other state $x'' \in \mathbb{X}^0$ in a finite number of steps. Corollary 4.6 then says that dissipativity is a necessary and sufficient condition for optimal steady-state operation if system (2.1) is weakly reversible. In case that the system is not weakly reversible, Theorem 4.5 only provides a partial converse result in the sense that dissipativity can only be ensured on a subset of $\mathbb{Z}^0$.

Remark 4.7. *The results of Theorem 4.5 are still valid for a slightly different definition of $\mathcal{Z}_M$. Namely, in (4.6), the sets $\mathcal{C}_M$ and $\mathcal{R}_M$ can also be defined by replacing x^* in (4.4) and (4.5), respectively, with any other state $y \in \mathbb{X}$.*

Necessity of dissipativity under a local controllability condition

In this section, we show that a *local* controllability condition is enough for establishing dissipativity on $\mathbb{Z}^0$ in case that system (2.1) is not only optimally operated at steady-state but uniformly suboptimally operated off steady-state, which we define as a slight extension of Definition 2.6.

Definition 4.8. *The system* (2.1) *is* uniformly suboptimally operated off steady-state, *if it is suboptimally operated off steady-state and in addition for each* δ, *there exists* $\bar{t} \in \mathbb{I}_{\geq 1}$ *such that for each solution satisfying* $(x(t), u(t)) \in \mathbb{Z}$ *for all* $t \in \mathbb{I}_{\geq 0}$ *at least one of the following two conditions holds:*

$$\sum_{t=0}^{T-1} \frac{\ell(x(t), u(t))}{T} \geq \ell(x^*, u^*) \quad \text{for all } T \in \mathbb{I}_{\geq \bar{t}}, \tag{4.10a}$$

$$|x(s) - x^*| \leq \delta \quad \text{for some } s \in \mathbb{I}_{[1, \bar{t}]}. \tag{4.10b}$$

Remark 4.9. *In Definition 4.8, uniformity is with respect to all initial conditions and feasible sequences. Namely, each feasible sequence either passes by arbitrarily close at the optimal steady-state within the finite time interval* $[1, \bar{t}]$ *(which only depends on the distance* δ *from* x^* *but not on the specific sequence) or has a transient performance greater than or equal to the optimal steady-state cost for all* $T \in \mathbb{I}_{\geq \bar{t}}$.

Before establishing necessity of dissipativity on $\mathbb{Z}^0$ for uniform suboptimal operation off steady-state, we slightly extend Theorem 2.9 and show that strict dissipativity is sufficient for uniform suboptimal operation off steady-state in case that the storage function λ is bounded on $\mathbb{X}^0$.

Theorem 4.10. *Suppose that system* (2.1) *is strictly dissipative on* $\mathbb{Z}^0$ *with respect to the supply rate* $s(x, u) = \ell(x, u) - \ell(x^*, u^*)$ *and with a storage function* λ *which is bounded on* $\mathbb{X}^0$. *Then the system* (2.1) *is uniformly suboptimally operated off steady-state.*

Proof. Sufficiency of strict dissipativity for suboptimal operation off steady-state was shown in Theorem 2.9. Hence it remains to show that (4.10a) or (4.10b) is satisfied for each feasible solution of system (2.1). From the strict dissipation inequality (2.12) and the fact that λ is bounded on $\mathbb{X}^0$, it follows that for each feasible solution and each $T \in \mathbb{I}_{\geq 0}$

$$-c := -2 \sup_{x \in \mathbb{X}^0} |\lambda(x)| \leq \lambda(x(T)) - \lambda(x(0)) \leq \sum_{t=0}^{T-1} \big(\ell(x(t), u(t)) - \ell(x^*, u^*) - \rho(|x(t) - x^*|)\big).$$

Let $\delta > 0$ be arbitrary but fixed, and define $\bar{t} := \lceil c/\rho(\delta) \rceil + 1$. Then, from the above inequality it directly follows that either $\sum_{t=0}^{T-1} \ell(x(t), u(t)) \geq T\ell(x^*, u^*)$ for all $T \geq \bar{t}$ (and hence also (4.10a) is satisfied), or $|x(s) - x^*| \leq \delta$ for at least two time instants $s \in \mathbb{I}_{[0, \bar{t}]}$ and hence for at least one time instant $s \in \mathbb{I}_{[1, \bar{t}]}$, i.e., (4.10b) holds. Hence system (2.1) is uniformly suboptimally operated off steady-state according to Definition 4.8. $\quad\square$

In order to examine necessity of dissipativity on $\mathbb{Z}^0$ for uniform suboptimal operation off steady-state, we need the following definition of local controllability, as can, e.g., be found in (Sontag, 1998, Section 3.7).

Definition 4.11 (Local controllability). *System* (2.1) *is* locally controllable *at* x^* *in* τ *steps if for every* $\hat{\varepsilon} > 0$, *there exists a* $\hat{\delta} > 0$ *such that for each pair of states* $y', y'' \in B_{\hat{\delta}}(x^*) \cap \mathbb{Z}$, *there exist an input sequence* $\boldsymbol{u}' : \mathbb{I}_{[0,\tau-1]} \to \mathbb{U}$ *and corresponding state sequence* $\boldsymbol{x}' : \mathbb{I}_{[0,\tau]} \to$ $\mathbb{X}$ *with* $x'(0) = y'$, $x'(\tau) = y''$, *and* $x'(t+1) = f(x'(t), u'(t))$ *and* $(x'(t), u'(t)) \in B_{\hat{\varepsilon}}(x^*, u^*) \cap \mathbb{Z}$ *for all* $t \in \mathbb{I}_{[0,\tau-1]}$.

In case that $(x^*, u^*) \in \text{int}(\mathbb{Z})$, there exists some $\hat{\varepsilon} > 0$ such that $B_{\varepsilon}(x^*, u^*) \cap \mathbb{Z} = B_{\varepsilon}(x^*, u^*)$ for all $0 \leq \varepsilon \leq \hat{\varepsilon}$. Then, in case that f is continuously differentiable, local controllability of system (2.1) in[2] n steps follows if the linearization of (2.1) at (x^*, u^*) is controllable (Sontag, 1998, Section 3.7). While the converse is not necessarily true, note that the latter is an easily verifiable condition. We can now state the following result.

Theorem 4.12. *Suppose that system* (2.1) *is uniformly suboptimally operated off steady-state and locally controllable at* x^* *in* τ *steps for some* $\tau \in \mathbb{I}_{\geq 0}$, *and that* ℓ *is locally bounded and bounded from below on*[3] $\mathbb{Z}^0$. *Then, system* (2.1) *is dissipative on* $\mathbb{Z}^0$ *with respect to the supply rate* $s(x, u) = \ell(x, u) - \ell(x^*, u^*)$ *and with a storage function* λ *which is bounded on* $\mathbb{X}^0$.

Proof. As was the case for Theorem 4.5, the proof of Theorem 4.12 will be done by contradiction. For simplicity and without loss of generality, we assume again in the following that $\ell(x^*, u^*) = 0$. Now assume for contradiction that the system is uniformly suboptimally operated off steady-state, but it is not dissipative on $\mathbb{Z}^0$ with λ bounded on $\mathbb{X}^0$. By Theorem 4.2 and Remark 4.3, this is equivalent to the fact that the available storage is unbounded on $\mathbb{X}^0$, and hence for each $r \geq 0$ there exists some $y \in \mathbb{X}^0$ such that

$$\inf_{\substack{T \geq 0 \\ z(0)=y \\ z(t+1)=f(z(t),v(t)) \\ (z(t),v(t))\in\mathbb{Z}^0 \ \forall t\in\mathbb{I}_{\geq 0}}} \sum_{t=0}^{T-1} \ell(z(t), v(t)) \leq -r. \tag{4.11}$$

This means that for each $r \geq 0$, there exist some $y \in \mathbb{X}^0$ and a state/input sequence pair $(\boldsymbol{x}_r, \boldsymbol{u}_r)$ together with a time instant $T_r \in \mathbb{I}_{\geq 0}$ such that $x_r(0) = y$, $(x_r(t), u_r(t)) \in \mathbb{Z}^0$ for all $t \in \mathbb{I}_{\geq 0}$ and (4.8) is satisfied. Note that as ℓ is bounded from below on $\mathbb{Z}^0$, it follows that $T_r \to \infty$ as $r \to \infty$. Now let $\hat{\varepsilon} > 0$ be arbitrary but fixed. As the system is uniformly suboptimally operated off steady-state, there exists $\bar{t} \in \mathbb{I}_{\geq 1}$ such that for all feasible sequences at least one of the conditions (4.10a)–(4.10b) is satisfied with $\delta = \hat{\delta}$, where $\hat{\delta}$ is the value from Definition 4.11 corresponding to $\hat{\varepsilon}$. Now define c as

$$c := \max\Big\{\tau \sup_{(x,u)\in B_{\hat{\varepsilon}}(x^*,u^*)\cap\mathbb{Z}^0} \ell(x,u), -\bar{t} \inf_{(x,u)\in\mathbb{Z}^0} \ell(x,u)\Big\}. \tag{4.12}$$

Note that $c < \infty$ as ℓ is assumed to be locally bounded (and hence bounded on the compact set $B_{\hat{\varepsilon}}(x^*, u^*)$) and bounded from below on $\mathbb{Z}^0$. Now consider some $r \geq 1 + 3c$ and note that, in this case, $T_r \geq 3\bar{t} + 1$ as $-r < 3\bar{t}\inf_{(x,u)\in\mathbb{Z}^0} \ell(x,u)$. Hence, due to

---

[2]Recall that n is the dimension of system (2.1), i.e., $x(t) \in \mathbb{R}^n$ for all $t \in \mathbb{I}_{\geq 0}$.

[3]We note that the first condition on ℓ (i.e., local boundedness) implies the second one (i.e., boundedness from below on $\mathbb{Z}^0$) if $\mathbb{Z}^0$ is compact. This is, e.g., the case if Assumptions 2.1 and 2.6 are satisfied, i.e., if f is continuous and $\mathbb{Z}$ is compact. Furthermore, note that a sufficient condition for local boundedness of ℓ is continuity, i.e., Assumption 2.3.

uniform suboptimal operation off steady-state, we conclude that $|x_r(s_1) - x^*| \leq \hat{\delta}$ for some $s_1 \in \mathbb{I}_{[1,\bar{t}]}$. Furthermore, as $\sum_{t=0}^{s_1-1} \ell(x_r(t), u_r(t)) \geq s_1 \inf_{(x,u) \in \mathbb{Z}^0} \ell(x, u) \geq -c$ by definition of c in (4.12), we have

$$\sum_{t=s_1}^{T_r-1} \ell(x_r(t), u_r(t)) \leq -(1 + 2c) \tag{4.13}$$

and $T_r - s_1 \geq 2\bar{t} + 1$ as $s_1 \leq \bar{t}$. We can now apply the above argument to the shifted sequence $x'_r(t) := x_r(t+s_1)$ and conclude by uniform suboptimal operation off steady-state that $|x'_r(s_2) - x^*| = |x_r(s_1 + s_2) - x^*| \leq \hat{\delta}$ for some $s_2 \in \mathbb{I}_{[1,\bar{t}]}$. Furthermore,

$$\sum_{t=s_1+s_2}^{T_r-1} \ell(x_r(t), u_r(t)) \leq -(1 + c)$$

by definition of c in (4.12), and $T_r - s_1 - s_2 \geq \bar{t} + 1$ as $s_2 \leq \bar{t}$. Repeating again the above argument, we conclude that $|x_r(s_1 + s_2 + s_3) - x^*| \leq \hat{\delta}$ for some $s_3 \in \mathbb{I}_{[1,\bar{t}]}$. We can now distinguish two different cases. Either we have

$$\sum_{t=s_1+s_2+s_3}^{T_r-1} \ell(x_r(t), u_r(t)) \geq -c, \tag{4.14}$$

or (4.14) does not hold, in which case the definition of c in (4.12) implies that $T_r - (s_1 + s_2 + s_3) > \bar{t}$. In the latter case, we can apply the above argument recursively to obtain time instances s_i, $i \in \mathbb{I}_{\geq 4}$, with $|x_r(s_1 + \cdots + s_i) - x^*| \leq \hat{\delta}$ until

$$\sum_{t=s_1+\cdots+s_j}^{T_r-1} \ell(x_r(t), u_r(t)) \geq -c, \tag{4.15}$$

for some $j \in \mathbb{I}_{\geq 4}$. Note that $j \leq T_r - \bar{t}$, as (4.15) is fulfilled as soon as $s_1 + \cdots + s_j \geq T_r - \bar{t}$ due to the definition of c in (4.12) and $s_1 + \cdots + s_j \geq j$.

Summarizing the above, we have proven that both $|x_r(s_1) - x^*| \leq \hat{\delta}$ and $|x_r(s_1 + \cdots + s_j) - x^*| \leq \hat{\delta}$, and

$$\sum_{k=s_1}^{s_1+\cdots+s_j-1} \ell(x_r(k), u_r(k)) \stackrel{(4.13),(4.15)}{\leq} -(1 + c). \tag{4.16}$$

Hence, by local controllability at the optimal steady-state (x^*, u^*) in τ steps, there exists a state/input sequence pair $(\boldsymbol{x}', \boldsymbol{u}')$ satisfying $x'(0) = x_r(s_1 + \cdots + s_j)$, $x'(\tau) = x_r(s_1)$, and $(x'(t), u'(t)) \in B_{\bar{\varepsilon}}(x^*, u^*) \cap \mathbb{Z}$ for all $t \in \mathbb{I}_{[0,\tau]}$. Furthermore, the definition of $\mathbb{Z}^0$ in (2.13) implies that also $(x'(t), u'(t)) \in B_{\bar{\varepsilon}}(x^*, u^*) \cap \mathbb{Z}^0$ for all $t \in \mathbb{I}_{[0,\tau]}$. By definition of c in (4.12) we have

$$\sum_{t=0}^{\tau-1} \ell(x'(t), u'(t)) \leq c. \tag{4.17}$$

Now define the following input sequence:

$$\hat{u}\big(k(s_2 + \cdots + s_j + \tau) + i\big) = \begin{cases} u_r(s_1 + i) & k \in \mathbb{I}_{\geq 0}, i \in \mathbb{I}_{[0,s_2+\cdots+s_j-1]} \\ u'(i) & k \in \mathbb{I}_{\geq 0}, i \in \mathbb{I}_{[s_2+\cdots+s_j,s_2+\cdots+s_j+\tau-1]} \end{cases} \tag{4.18}$$

which results in a cyclic state sequence with $\hat{x}(k(s_2 + \cdots + s_j + \tau)) = x_r(s_1)$ for all $k \in \mathbb{I}_{\geq 0}$. By construction, this state/input sequence pair fulfills $(\hat{x}(t), \hat{u}(t)) \in \mathbb{Z}^0$ for all $t \in \mathbb{I}_{\geq 0}$. Furthermore, we obtain for all $k \in \mathbb{I}_{\geq 0}$:

$$\sum_{i=0}^{s_2+\cdots+s_j+\tau-1} \ell\Big(\hat{x}\big(k(s_2 + \cdots + s_j + \tau - 1) + i\big), \hat{u}\big(k(s_2 + \cdots + s_j + \tau - 1) + i\big)\Big) \overset{(4.16)-(4.18)}{\leq} -1.$$
$$\tag{4.19}$$

However, this implies that

$$\liminf_{T \to \infty} \sum_{t=0}^{T-1} \frac{\ell(\hat{x}(t), \hat{u}(t))}{T} \overset{(4.18)}{=} \frac{1}{s_2 + \cdots + s_j + \tau} \sum_{i=0}^{s_2+\cdots+s_j+\tau-1} \ell(\hat{x}(i), \hat{u}(i))$$
$$\overset{(4.19)}{\leq} -\frac{1}{s_2 + \cdots + s_j + \tau} < 0$$

which contradicts (2.9) and hence the assumption of uniform suboptimal operation off steady-state. Therefore, we conclude that the system (2.1) is dissipative on $\mathbb{Z}^0$ with respect to the supply rate $s(x,u) = \ell(x,u) - \ell(x^*,u^*)$ and with a storage function λ which is bounded on $\mathbb{X}^0$. $\qquad \square$

Discussion and summary

Going back to Example 4.4, we can now interpret why the considered system was not dissipative despite being optimally operated at steady-state. Namely, from (4.3), it immediately follows that $\mathcal{C}_M = \mathbb{X}$ and $\mathcal{R}_M = \{0\}$ for each $M \in \mathbb{I}_{\geq 1}$, and hence $\mathcal{Z}_M = \{0\}$. Thus, Theorem 4.5 does not yield any nontrivial result for Example 1. Furthermore, while system (4.3) is uniformly suboptimally operated off steady-state and ℓ is bounded on $\mathbb{Z}$, Theorem 4.12 cannot be applied as the system is not locally controllable at $x^* = 0$ (and hence in particular also its linearization at $x^* = 0$ is not controllable). Hence the lack of dissipativity of system (4.3) results from a lack of controllability.

Note that while in Theorem 4.10 strict dissipativity was needed in order to prove suboptimal operation off steady-state, in the converse result (Theorem 4.12) only dissipativity, but not strict dissipativity, could be established. In fact, if one wanted to establish strict dissipativity by contradiction as in the proof of Theorem 4.12, one would obtain that for each $r \geq 0$ and each $\rho \in \mathcal{K}_\infty$, there exists some $y \in \mathbb{X}^0$ such that (4.11) holds with additional term $-\rho(|z(k) - x^*|)$ in the sum of the left hand side. From here, however, one cannot establish a contradiction to the system being uniformly suboptimally operated off steady-state, except for the case $\rho \equiv 0$. Hence, only dissipativity and not strict dissipativity could be established.

Furthermore, as was discussed in Section 2.2.2, in case that the system has to fulfill additional asymptotic average constraints, dissipativity with respect to the (relaxed) supply rate $\bar{s}(x,u) = \ell(x,u) - \ell(x^*,u^*) + \bar{\lambda}^T h(x,u)$ for some (arbitrary) $\bar{\lambda} \in \mathbb{R}^p_{\geq 0}$ is sufficient for optimal steady-state operation. For this setting, necessity of dissipativity cannot be established as shown in the proofs of Theorems 4.5 and 4.12, as the input sequences $\hat{\boldsymbol{u}}$ constructed in (4.9) and (4.18), respectively, together with the corresponding state sequences $\hat{\boldsymbol{x}}$, are not necessarily guaranteed to fulfill the average constraints and hence cannot be used to induce a contradiction. Therefore, necessity of dissipativity for optimal steady-state operation in the context of economic MPC with average constraints is currently still an open research question.

To summarize, in this section we have shown that dissipativity is in general not necessary for a system to be optimally operated at steady-state. Nevertheless, we have shown that necessity of dissipativity for optimal steady-state operation and uniform suboptimal operation off steady-state, respectively, can be established under rather mild controllability conditions on the system. Hence in conclusion, we find that dissipativity is an adequate and precise (i.e., not too conservative) characterization of optimal steady-state operation. Furthermore, as already mentioned in Remark 2.12, the results presented in this section are also of importance in the context of stability and convergence analysis of economic MPC, as there typically the existence of a storage function λ is sufficient but λ does not have to be known (see, e.g., Theorem 2.10 and Angeli et al. (2012); Grüne (2013)). Theorems 4.5 and 4.12 guarantee existence of such a storage function λ based on certain dynamic properties (controllability and optimal steady-state operation) of the considered system.

4.1.2 Robustness of dissipativity under changing constraint sets

The supply rate of interest in economic MPC, $s(x, u) = \ell(x, u) - \ell(x^*, u^*)$, depends on the state and input constraints which act on the system, namely through the optimal steady-state (x^*, u^*). In this section, we examine what happens if the constraints are changed, and hence also the supply rate is altered. We first show with a simple motivating example that dissipativity of system (2.1) may be lost even for arbitrarily small changes in the constraint set. We then provide a robustness analysis of the considered dissipativity property, and show that further results are possible if a certain convexity assumption is satisfied. For a brief overview on the necessary background in nonlinear programming and the notation used in this section, the reader is referred to Appendix B.

While the focus of this section is to examine robustness of dissipativity in light of both small and large changes in the constraints, we note that some of the results can also be extended to uncertainties in the cost function and the system dynamics, as well as to economic MPC settings with average constraints (see Remark 4.19 for more details).

A motivating example

Example 4.13. *Consider the scalar system*

$$x(t + 1) = x(t)u(t) \qquad (4.20)$$

with state and input constraint set $\mathbb{Z} = \mathbb{X} \times \mathbb{U} = [-5, 5]^2$. The set of all feasible state/input equilibrium pairs is computed according to (2.3) as $S = \{(x, u) : x = 0, u \in \mathbb{U}\} \cup \{(x, u) : x \in \mathbb{X}, u = 1\}$. The stage cost ℓ is given by $\ell(x, u) = (x-1)^2 + \delta(u - \bar{u})^2$ for some $\bar{u} \in \mathbb{R}$ and $0 < \delta < 1/(\bar{u} - 1)^2$ (respectively, $\delta > 0$ in case that $\bar{u} = 1$). Straightforward calculations yield that the set of optimal state/input equilibrium pairs (x^*, u^*) as defined in (2.7) is

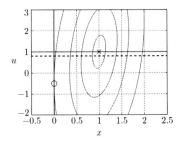

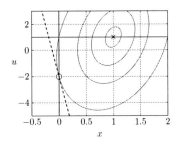

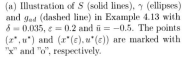

(a) Illustration of S (solid lines), γ (ellipses) and g_{ad} (dashed line) in Example 4.13 with $\delta = 0.035$, $\varepsilon = 0.2$ and $\bar{u} = -0.5$. The points (x^*, u^*) and $(x^*(\varepsilon), u^*(\varepsilon))$ are marked with "x" and "o", respectively.

(b) Illustration of S (solid lines), γ (ellipses) and $\hat{g}_{ad}$ (dashed line) in Example 4.22 with $\delta = 0.035$ and $\bar{u} = -0.5$. The points (x^*, u^*) and (x^*_{ad}, u^*_{ad}) are marked with "x" and "o", respectively.

Figure 4.1: Example 4.13 (left) and Example 4.22 (right).

$S^* = \{(1,1)\}$, with $\ell(x^*, u^*) = \delta(1 - \bar{u})^2$. One can show that the system is dissipative on $\mathbb{Z}$ with respect to the supply rate $s(x, u) = \ell(x, u) - \ell(x^*, u^*)$ and with storage function $\lambda(x) = 2\delta(1 - \bar{u})x$, which means that (x^*, u^*) minimizes the function γ on $\mathbb{Z}$, where

$$\gamma(x, u) := \ell(x, u) - \ell(x^*, u^*) + \lambda(x) - \lambda(f(x, u)). \tag{4.21}$$

The set S as well as $\text{lev}(\gamma)$, i.e., level sets of γ (with $\delta = 0.035$ and $\bar{u} = -0.5$), are depicted in Figure 4.1(a).

Now consider an additional constraint of the form $g_{ad}(x, u; \varepsilon) = u - (1 - \varepsilon) \leq 0$ for some $\varepsilon \leq 1$. If $\varepsilon \leq 0$, then (x^*, u^*) is still feasible and hence the system is still dissipative with the same storage function and supply rate. On the other hand, if $\varepsilon > 0$, the optimal steady-state $(x^*, u^*) = (1, 1)$ becomes infeasible. In this case, the new set $S^*(\varepsilon)$ of optimal feasible steady-states $(x^*(\varepsilon), u^*(\varepsilon))$ can be calculated as $S^*(\varepsilon) = \{(0, \min\{\bar{u}, 1 - \varepsilon\})\}$, i.e., $S^*(\varepsilon) \neq S^*$, and $\ell(x^*(\varepsilon), u^*(\varepsilon)) = 1 + \delta(\min\{\bar{u}, 1 - \varepsilon\} - \bar{u})^2$. We will show that for certain values of δ and $\bar{u}$, there does not exist any storage function $\lambda(x; \varepsilon)$ such that system (4.20) is dissipative with respect to the new supply rate $s(x, u; \varepsilon) = \ell(x, u) - \ell(x^*(\varepsilon), u^*(\varepsilon))$ for all (x, u) in the new constraint set. By Theorem 4.2, this can be done by showing that the available storage S_a is unbounded for some x. Namely, consider for $x_0 = 1$ the following feasible input sequence $\boldsymbol{u} = \{-1/3, -3, -1/3, -3, \dots\}$ with corresponding state sequence $\boldsymbol{x} = \{1, -1/3, 1, -1/3, \dots\}$. Along this sequence, we obtain that

$$\limsup_{T \to \infty} \sum_{t=0}^{T-1} -s(x(t), u(t); \varepsilon) = T\left(\ell(x^*(\varepsilon), u^*(\varepsilon)) - \frac{(4/3)^2}{2} - \frac{\delta((\bar{u} + 1/3)^2 + (\bar{u} + 3)^2)}{2}\right)$$

$$=: Ta.$$

Straightforward calculations yield that for each $0 < \varepsilon \leq 1$, we have $a > 0$ if

$$\delta < \frac{1}{(30 + 18\min\{\bar{u}, 1\})\bar{u} + 41 - 9\min\{\bar{u}, 1\}^2}. \tag{4.22}$$

For example, for $\bar{u} = -0.5$ as in Figure 4.1(a), (4.22) yields $\delta < 0.0354$. Then, due to $a > 0$ and the definition of the available storage in (4.1), we have $S_a(1) = \infty$. Thus, by Theorem 4.2, there does not exist any storage function $\lambda(x; \varepsilon)$ such that the system (4.20) is dissipative with respect to the supply rate $s(x, u; \varepsilon) = \ell(x, u) - \ell(x^*(\varepsilon), u^*(\varepsilon))$. This means that an arbitrarily small change in the additional constraints (from $\varepsilon = 0$ to ε slightly greater zero) can destroy the considered dissipativity property and hence also optimal operation of system (4.20) at steady-state.

Robustness of dissipativity with respect to changing supply rate

In this Section, we give conditions under which the situation of Example 4.13 cannot occur. In particular, they are such that small changes in the constraints result in small changes in the optimal steady-state and hence in the supply rate, in which case robustness of the considered dissipativity property can be established. To this end, in the following we assume that the state and input constraint set $\mathbb{Z}$ in (2.2) is given in the form of inequality constraints, which depend on additional parameters $\varepsilon \in \mathbb{R}^s$, i.e.,

$$\mathbb{Z}(\varepsilon) := \{(x, u) : g(x, u; \varepsilon) \le 0\} \qquad (4.23)$$

for some function $g : \mathbb{R}^n \times \mathbb{R}^m \times \mathbb{R}^s \to \mathbb{R}^r$. Similar to Assumption 2.6, we make the following compactness assumption on the sets $\mathbb{Z}(\varepsilon)$.

Assumption 4.1. *There exists some $\varepsilon_{\max} > 0$ and some compact set $\mathbb{Z}_{\max}$ such that for all $0 \le |\varepsilon| \le \varepsilon_{\max}$, the set $\mathbb{Z}(\varepsilon)$ is non-empty and $\mathbb{Z}(\varepsilon) \subseteq \mathbb{Z}_{\max}$.*

Remark 4.14. *For typical results in economic MPC such as optimal steady-state operation or stability (see, e.g., Sections 2.2 and 4.1.1), dissipativity on $\mathbb{Z}^0(\varepsilon)$ instead of $\mathbb{Z}(\varepsilon)$ would be enough, where $\mathbb{Z}^0(\varepsilon)$ is defined analogous to (2.13) with $\mathbb{Z}$ replaced by $\mathbb{Z}(\varepsilon)$. However, as discussed in Remark 2.12, $\mathbb{Z}^0(\varepsilon)$ might in general be difficult to compute. On the other hand, assuming that the state and input constraint set $\mathbb{Z}(\varepsilon)$ is given in terms of inequality constraints as in (4.23) is not a major restriction, which is why in this section we consider robustness of dissipativity on the set $\mathbb{Z}(\varepsilon)$. Nevertheless, the following results are also valid with $\mathbb{Z}(\varepsilon)$ replaced by $\mathbb{Z}^0(\varepsilon)$ (or any other set $\widetilde{Z}(\varepsilon)$), if $\mathbb{Z}^0(\varepsilon)$ (respectively, $\widetilde{Z}(\varepsilon)$) can be described by inequalities as in (4.23) and satisfies Assumption 4.1.*

The above definition of $\mathbb{Z}(\varepsilon)$ also includes the situation of Example 4.13, where additional constraints g_{ad} are imposed on the system; in this case, the function g in (4.23) consists of the inequality constraints describing the original constraint set $\mathbb{Z}$ and the additional constraints g_{ad}. The sets of feasible and optimal state/input equilibrium pairs can now be defined analogously to Chapter 2, i.e., $S(\varepsilon) := \{(x, u) \in \mathbb{Z}(\varepsilon) : x = f(x, u)\}$ and

$$S^*(\varepsilon) := \{(y, w) \in S(\varepsilon) : \ell(y, w) = \min_{(x,u) \in S(\varepsilon)} \ell(x, u)\}. \qquad (4.24)$$

In the following, we assume that $S(\varepsilon)$ (and hence also $S^*(\varepsilon)$) are non-empty for all $0 \le |\varepsilon| \le \varepsilon_{\max}$. As above, in the following $(x^*(\varepsilon), u^*(\varepsilon))$ denotes an arbitrary element of $S^*(\varepsilon)$. Note that $(x^*(\varepsilon), u^*(\varepsilon))$ is a global minimizer of the problem

$$\mathcal{P}_\ell[\varepsilon] := \mathcal{P}([x^T \ u^T]^T, \ell, x - f(x, u), g) \qquad (4.25)$$

as defined in equation (B.1) in Appendix B. The question we are interested in is under what circumstances there exists a storage function $\lambda(x;\varepsilon)$ such that, if the system (2.1) is dissipative with $\varepsilon = 0$, for changing ε it remains dissipative for all $(x,u) \in \mathbb{Z}(\varepsilon)$ with respect to the supply rate $s(x,u;\varepsilon) = \ell(x,u) - \ell(x^*(\varepsilon), u^*(\varepsilon))$. This means that the function

$$\gamma(x,u;\varepsilon) := \ell(x,u) - \ell(x^*(\varepsilon), u^*(\varepsilon)) + \lambda(x;\varepsilon) - \lambda(f(x,u);\varepsilon) \tag{4.26}$$

satisfies $\gamma(x,u;\varepsilon) \geq 0$ for all $(x,u) \in \mathbb{Z}(\varepsilon)$, i.e., $(x^*(\varepsilon), u^*(\varepsilon))$ is a global minimizer of the problem

$$\mathcal{P}_\gamma[\varepsilon] := \mathcal{P}([x^T \ u^T]^T, \gamma, 0, g) \tag{4.27}$$

as defined in (B.1) with $\gamma(x^*(\varepsilon), u^*(\varepsilon); \varepsilon) = 0$. As demonstrated in Example 4.13, even if the changes in ε are arbitrarily small, in general the system might lose its dissipativity property as soon as the optimal steady-state, and hence also the supply rate, is altered. The following Theorem shows under what conditions robustness of (strict) dissipativity with respect to small changes in ε can be guaranteed. To this end, we need the following assumption on the functions f, ℓ and g.

Assumption 4.2. *The functions f, ℓ and g are twice continuously differentiable in (x,u). Furthermore, g as well as its first and second derivatives with respect to (x,u) are continuous in ε.*

Theorem 4.15. *Suppose that Assumptions 4.1 and 4.2 hold and the following is satisfied:*

(i) For $\varepsilon = 0$, $S^*(0)$ is a singleton, i.e., $(x^*(0), u^*(0))$ is the unique optimal steady-state. Furthermore, $(x^*(0), u^*(0))$ is the unique global minimizer of problem $\mathcal{P}_\gamma[0]$, i.e., system (2.1) is strictly dissipative on $\mathbb{Z}(0)$ with respect to the supply rate $s(x,u;0) = \ell(x,u) - \ell(x^*(0), u^*(0))$, and the corresponding storage function $\lambda(x;0)$ is twice continuously differentiable in x.

(ii) The optimal steady-state $(x^*(0), u^*(0))$ is regular and satisfies the strong second order sufficiency condition (B.4) for problems $\mathcal{P}_\ell[0]$ and $\mathcal{P}_\gamma[0]$.

Then there exists $\bar{\varepsilon}$ with $0 < \bar{\varepsilon} \leq \varepsilon_{\max}$ such that for all $|\varepsilon| \leq \bar{\varepsilon}$ the system (2.1) is strictly dissipative on $\mathbb{Z}(\varepsilon)$ with respect to the supply rate $s(x,u;\varepsilon) = \ell(x,u) - \ell(x^*(\varepsilon), u^*(\varepsilon))$ and with storage function $\lambda(x;\varepsilon) := \lambda(x;0) + \tilde{\lambda}(\varepsilon)^T x$, where $\tilde{\lambda}(\varepsilon)$ is continuous in ε with $\tilde{\lambda}(0) = 0$.

Theorem 4.15 means that the storage function $\lambda(x;\varepsilon)$ can be modified continuously with changing parameters ε such that the dissipativity property of the system is maintained. In particular, the (in general nonlinear) storage function λ can be modified by an additional term $\tilde{\lambda}(\varepsilon)^T x$, which is linear in x, in order to still serve as a storage function for the system under changing constraints. We now discuss some properties and implications of Theorem 4.15, before proceeding with its proof.

Remark 4.16. *In view of Theorems 2.9 and 2.10, it follows from Theorem 4.15 that system (2.1) is robustly optimally operated at steady-state and robustly asymptotically stable with respect to small changes in the state and input constraints.*

Remark 4.17. *Note that $(x^*(0), u^*(0))$ being the unique global minimizer of problem $\mathcal{P}_\gamma[0]$ as required in condition (i) of Theorem 4.15 is slightly stronger than strict dissipativity of system (2.1) on $\mathbb{Z}(0)$ with respect to the supply rate $s(x, u; 0) = \ell(x, u) - \ell(x^*(0), u^*(0))$. Namely, as $\mathbb{Z}(0)$ is compact and γ is continuous, $(x^*(0), u^*(0))$ being the unique global minimizer of problem $\mathcal{P}_\gamma[0]$ implies[4] that there exists a function $\rho' \in \mathcal{K}_\infty$ such that $\gamma(x, u; 0) \geq \rho'(|(x - x^*(0), u - u^*(0))|)$ for all $(x, u) \in \mathbb{Z}(0)$, whereas strict dissipativity according to Definition 2.7 means that $\gamma(x, u; 0) \geq \rho(|x - x^*(0)|)$ for some $\rho \in \mathcal{K}_\infty$. The proof of Theorem 4.15 (see below) reveals that under the given hypotheses, also $(x^*(\varepsilon), u^*(\varepsilon))$ is the unique global minimizer of problem $\mathcal{P}_\gamma[\varepsilon]$ for all $|\varepsilon| \leq \bar{\varepsilon}$, and hence system (2.1) is strictly dissipative on $\mathbb{Z}(\varepsilon)$ with respect to the supply rate $s(x, u; \varepsilon) = \ell(x, u) - \ell(x^*(\varepsilon), u^*(\varepsilon))$.*

Proof of Theorem 4.15. The proof of Theorem 4.15 consists of two parts. First, the sensitivity analysis in nonlinear programming (Fiacco and Ishizuka, 1990; Jittorntrum, 1984; Robinson, 1980) is applied to conclude that for sufficiently small $|\varepsilon|$, there exists a steady-state $(x^*(\varepsilon), u^*(\varepsilon))$ which is continuous in ε and a strict local minimizer of problem $\mathcal{P}_\ell[\varepsilon]$. We then show that the storage function $\lambda(x; \varepsilon)$ can be modified continuously in ε such that $(x^*(\varepsilon), u^*(\varepsilon))$ is also a strict local minimizer of problem $\mathcal{P}_\gamma[\varepsilon]$. In the second part, we show that $(x^*(\varepsilon), u^*(\varepsilon))$ is not only a local but also the unique global minimizer of problems $\mathcal{P}_\ell[\varepsilon]$ and $\mathcal{P}_\gamma[\varepsilon]$, which implies that indeed $(x^*(\varepsilon), u^*(\varepsilon)) \in S^*(\varepsilon)$ according to the definition of $\mathcal{P}_\ell[\varepsilon]$ in (4.25), and that the system (2.1) is strictly dissipative on $\mathbb{Z}(\varepsilon)$ with respect to the supply rate $s(x, u; \varepsilon) = \ell(x, u) - \ell(x^*(\varepsilon), u^*(\varepsilon))$ according to the definition of $\mathcal{P}_\gamma[\varepsilon]$ in (4.27), as discussed in Remark 4.17.

Part 1: Let $h_s(x, u) := x - f(x, u)$ and $\Lambda(x, u; \varepsilon) := \lambda(x; \varepsilon) - \lambda(f(x, u); \varepsilon)$. As $(x^*(0), u^*(0))$ is regular and a strict minimizer of both problems $\mathcal{P}_\ell[0]$ and $\mathcal{P}_\gamma[0]$, by Proposition B.1 there exist unique Lagrange multipliers $\mu_\ell(0) \in \mathbb{R}^n$, $\nu_\ell(0) \in \mathbb{R}^r$ and $\nu_\gamma(0) \in \mathbb{R}^r$ such that the following is satisfied:

$$\nabla_{(x,u)}\ell(x^*(0), u^*(0)) + \mu_\ell(0)^T \nabla_{(x,u)} h_s(x^*(0), u^*(0)) + \nu_\ell(0)^T \nabla_{(x,u)} g(x^*(0), u^*(0); 0) = 0,$$
(4.28)

$$\nabla_{(x,u)}\ell(x^*(0), u^*(0)) + \nabla_{(x,u)}\Lambda(x^*(0), u^*(0); 0) + \nu_\gamma(0)^T \nabla_{(x,u)} g(x^*(0), u^*(0); 0) = 0. \quad (4.29)$$

Now consider the term $\nabla_{(x,u)}\Lambda(x^*(0), u^*(0); 0)$. We obtain

$$\begin{aligned}
&\nabla_{(x,u)}\Lambda(x^*(0), u^*(0); 0) \\
&= \nabla_x\lambda(x^*(0); 0) \begin{bmatrix} I_n & 0_{n\times m} \end{bmatrix} - \nabla_x\lambda(f(x^*(0), u^*(0)); 0)\nabla_{(x,u)}f(x^*(0), u^*(0)) \\
&= \nabla_x\lambda(x^*(0); 0)\Big(\begin{bmatrix} I_n & 0_{n\times m} \end{bmatrix} - \nabla_{(x,u)}f(x^*(0), u^*(0)) \Big) \\
&= \nabla_x\lambda(x^*(0); 0)\nabla_{(x,u)}h_s(x^*(0), u^*(0))
\end{aligned}$$
(4.30)

where the first equality follows from the chain rule, the second is due to the fact that $(x^*(0), u^*(0))$ is a steady-state, i.e., $f(x^*(0), u^*(0)) = x^*(0)$, and the third follows from the definition of h_s. Comparing (4.28) with (4.29) and using (4.30) as well as the fact that the Lagrange multipliers are unique, we obtain

$$\nabla_x\lambda(x^*(0); 0) = \mu_\ell(0)^T, \quad \nu_\gamma(0) = \nu_\ell(0). \quad (4.31)$$

---

[4]This follows, for example, from Lemma 4.3 in (Khalil, 2002).

Next, as $(x^*(0), u^*(0))$ is assumed to fulfill the strong second order sufficiency condition (B.4) for problem $\mathcal{P}_\ell[0]$, from the sensitivity analysis in[5] (Jittorntrum, 1984, Theorem 2) (see also (Robinson, 1980, Section 4) and (Fiacco and Ishizuka, 1990, Theorem 5.2)) we obtain the result that there exists $0 < \varepsilon_1 \leq \varepsilon_{\max}$ such that for all $|\varepsilon| \leq \varepsilon_1$, the problem $\mathcal{P}_\ell[\varepsilon]$ has a unique local minimizer[6] $(x^*(\varepsilon), u^*(\varepsilon))$ which is regular and continuous in ε as well as the corresponding unique Lagrange multipliers $\mu_\ell(\varepsilon)$ and $\nu_\ell(\varepsilon)$.

Now consider the problem $\mathcal{P}_\gamma[\varepsilon]$ defined in (4.26)–(4.27), with $\lambda(x; \varepsilon)$ given by

$$\lambda(x; \varepsilon) := \lambda(x; 0) + \tilde{\lambda}(\varepsilon)^T x \tag{4.32}$$

with

$$\tilde{\lambda}(\varepsilon)^T := \mu_\ell(\varepsilon)^T - \nabla_x \lambda(x^*(\varepsilon); 0). \tag{4.33}$$

Note that as $x^*(\varepsilon)$ is continuous in ε, the same holds true for $\nabla_x \lambda(x^*(\varepsilon); 0)$. As furthermore also $\mu_\ell(\varepsilon)$ is continuous in ε, it follows that $\tilde{\lambda}(\varepsilon)$ is continuous in ε with $\tilde{\lambda}(0) = 0$ due to (4.31) and (4.33). But this implies that also the function $\lambda(x; \varepsilon)$ and hence also $\gamma(x, u; \varepsilon)$ are continuous in ε. Moreover, from the above and Assumption 4.2 it follows that $\nabla^2_{(x,u)} \gamma(x^*(\varepsilon), u^*(\varepsilon); \varepsilon)$ is continuous in ε.

As next step, we want to show that $(x^*(\varepsilon), u^*(\varepsilon))$ is not only a strict local minimizer of problem $\mathcal{P}_\ell[\varepsilon]$, but also of problem $\mathcal{P}_\gamma[\varepsilon]$. By Proposition B.2, this can be concluded if $(x^*(\varepsilon), u^*(\varepsilon))$ satisfies both the KKT conditions (B.2)–(B.3) (with some $\nu_\gamma(\varepsilon)$) and the strong second order sufficiency condition (B.4) for problem $\mathcal{P}_\gamma[\varepsilon]$. We first verify the KKT conditions. Taking $\nu_\gamma(\varepsilon) = \nu_\ell(\varepsilon)$, (B.3) is immediately satisfied. Furthermore, (B.2) equals (4.29) with 0 replaced by ε. Using (4.30) with 0 replaced by ε as well as (4.32) and (4.33), we obtain

$$\nabla_{(x,u)}\ell(x^*(\varepsilon), u^*(\varepsilon)) + \nabla_{(x,u)}\Lambda(x^*(\varepsilon), u^*(\varepsilon); \varepsilon) + \nu_\gamma(\varepsilon)^T \nabla_{(x,u)} g(x^*(\varepsilon), u^*(\varepsilon); \varepsilon)$$

$$\overset{(4.30)}{=} \nabla_{(x,u)}\ell(x^*(\varepsilon), u^*(\varepsilon)) + \nabla_x \lambda(x^*(\varepsilon); \varepsilon)\nabla_{(x,u)} h_s(x^*(\varepsilon), u^*(\varepsilon))$$

$$\qquad + \nu_\gamma(\varepsilon)^T \nabla_{(x,u)} g(x^*(\varepsilon), u^*(\varepsilon); \varepsilon)$$

$$\overset{(4.32),(4.33)}{=} \nabla_{(x,u)}\ell(x^*(\varepsilon), u^*(\varepsilon)) + \mu_\ell(\varepsilon)^T \nabla_{(x,u)} h_s(x^*(\varepsilon), u^*(\varepsilon)) + \nu_\ell(\varepsilon)^T \nabla_{(x,u)} g(x^*(\varepsilon), u^*(\varepsilon); \varepsilon)$$

$$= 0.$$

The last equality follows from the fact that by Proposition B.1, equation (4.28) is satisfied with 0 replaced by ε as $(x^*(\varepsilon), u^*(\varepsilon))$ is a strict local minimizer of problem $\mathcal{P}_\ell[\varepsilon]$. Hence $(x^*(\varepsilon), u^*(\varepsilon))$ satisfies the KKT conditions for problem $\mathcal{P}_\gamma[\varepsilon]$.

Next, we show that $(x^*(\varepsilon), u^*(\varepsilon))$ also satisfies the strong second order sufficiency condition (B.4) for problem $\mathcal{P}_\gamma[\varepsilon]$, which reads

$$w^T \left(\nabla^2_{(x,u)} \gamma(x^*(\varepsilon), u^*(\varepsilon); \varepsilon) + \sum_{j=1}^r \nu_{\gamma,j}(\varepsilon) \nabla^2_{(x,u)} g_j(x^*(\varepsilon), u^*(\varepsilon); \varepsilon) \right) w > 0 \tag{4.34}$$

---

[5] In various sensitivity results like (Jittorntrum, 1984) it is assumed that the function g is twice continuously differentiable in (x, u, ε). However, differentiability with respect to ε can be relaxed to Assumption 4.2, if only continuity (but not differentiability) of the locally optimal solution and the corresponding Lagrange multipliers with respect to ε shall be established (compare (Robinson, 1980) and (Fiacco and Ishizuka, 1990, Section 5)), which is what we need here.

[6] With a slight abuse of notation, we already denote this local minimizer by $(x^*(\varepsilon), u^*(\varepsilon))$, although we earlier reserved this notation for a global minimizer of problem $\mathcal{P}_\ell[\varepsilon]$. We will show later that $(x^*(\varepsilon), u^*(\varepsilon))$ is not only a local but indeed a global minimizer of problem $\mathcal{P}_\ell[\varepsilon]$.

for all $w \neq 0$ such that $\nabla_{(x,u)} g_j(x^*(\varepsilon), u^*(\varepsilon); \varepsilon) w = 0$ for all $j \in A([x^*(\varepsilon)^T \ u^*(\varepsilon)^T]^T)$ such that $\nu_{\gamma,j}(\varepsilon) > 0$. Note that as $\nu_\gamma(\varepsilon) = \nu_\ell(\varepsilon)$ is continuous in ε, for sufficiently small $|\varepsilon|$ it holds that $\nu_{\gamma,j}(\varepsilon) > 0$ for all j such that $\nu_{\gamma,j}(0) > 0$. But then, due to continuity reasons, (4.34) is satisfied for sufficiently small $|\varepsilon|$ as it is satisfied by assumption for $\varepsilon = 0$. Namely, it this were not the case, then there would exist an infinite sequence $\{(x^*(\varepsilon_k), u^*(\varepsilon_k))\}$ with $\varepsilon_k \to 0$ and a corresponding sequence $\{w_k\}$ with $|w_k| = 1$ such that

$$w_k^T \left(\nabla^2_{(x,u)} \gamma(x^*(\varepsilon_k), u^*(\varepsilon_k); \varepsilon_k) + \sum_{j=1}^{r} \nu_{\gamma,j}(\varepsilon_k) \nabla^2_{(x,u)} g_j(x^*(\varepsilon_k), u^*(\varepsilon_k); \varepsilon_k) \right) w_k \leq 0$$

and $\nabla_{(x,u)} g_j(x^*(\varepsilon_k), u^*(\varepsilon_k); \varepsilon_k) w_k = 0$ for all $j \in A([x^*(\varepsilon_k)^T \ u^*(\varepsilon_k)^T]^T)$ such that $\nu_{\gamma,j}(\varepsilon_k) > 0$. As the sequence w_k is bounded, it has a convergent subsequence. Taking the limit over such a subsequence results in a contradiction to (4.34) with $\varepsilon = 0$, as $\nabla^2_{(x,u)} \gamma(x^*(\varepsilon), u^*(\varepsilon); \varepsilon)$, $\nu_{\gamma,j}(\varepsilon)$, $\nabla_{(x,u)} g_j(x^*(\varepsilon), u^*(\varepsilon); \varepsilon)$ and $\nabla^2_{(x,u)} g_j(x^*(\varepsilon), u^*(\varepsilon); \varepsilon)$ are continuous in ε.

Part 2: Summarizing the above, we have shown that for sufficiently small $|\varepsilon|$, there exists $(x^*(\varepsilon), u^*(\varepsilon))$ which is continuous in ε and is a strict local minimizer of both problems $\mathcal{P}_\ell[\varepsilon]$ and $\mathcal{P}_\gamma[\varepsilon]$. What remains to show is that for sufficiently small $|\varepsilon|$, $(x^*(\varepsilon), u^*(\varepsilon))$ is also a global minimizer of both problems $\mathcal{P}_\ell[\varepsilon]$ and $\mathcal{P}_\gamma[\varepsilon]$. To this end, consider the following. As $(x^*(\varepsilon), u^*(\varepsilon))$ is a strict local minimizer of problem $\mathcal{P}_\gamma[\varepsilon]$ and furthermore g, γ and $(x^*(\varepsilon), u^*(\varepsilon))$ are continuous in ε, there exists $\delta > 0$ such that $(x^*(\varepsilon), u^*(\varepsilon))$ is a strict minimizer of γ on the set $\mathbb{Z}(\varepsilon) \cap B_\delta(x^*(0), u^*(0))$. Furthermore, according to Assumption (i) of Theorem 4.15, $(x^*(0), u^*(0))$ is a strict global minimizer of problem $\mathcal{P}_\gamma[0]$, i.e., $(x^*(0), u^*(0))$ uniquely minimizes $\gamma(x, u; 0)$ over $\mathbb{Z}(0)$. Hence $\gamma(x, u; 0) > 0$ for all $(x, u) \in \mathbb{Z}(0) \setminus B_\delta(x^*(0), u^*(0))$. But then, due to continuity of γ in (x, u, ε), it holds that also $\gamma(x, u; \varepsilon) > 0$ for all $(x, u) \in \mathcal{N}(\mathbb{Z}(0)) \setminus B_\delta(x^*(0), u^*(0))$ for each sufficiently small neighborhood $\mathcal{N}(\mathbb{Z}(0))$ of $\mathbb{Z}(0)$ and sufficiently small $|\varepsilon|$. For any such open neighborhood, let $g_{j,\min}(\varepsilon) := \min_{(x,u) \in \mathbb{Z}_{\max} \setminus \mathcal{N}(\mathbb{Z}(0))} g_j(x, u, \varepsilon)$, for all $j \in \mathbb{I}_{[1,r]}$. Note that $g_{j,\min}(\varepsilon)$ is well defined[7] by compactness of $\mathbb{Z}_{\max} \setminus \mathcal{N}(\mathbb{Z}(0))$ and continuity of g. Furthermore $g_{j,\min}(\varepsilon) > 0$ for sufficiently small $|\varepsilon|$ as g is continuous in ε and $g_{j,\min}(0) > 0$. But then, by definition of $\mathbb{Z}(\varepsilon)$ in (4.23) and the fact that $\mathbb{Z}(\varepsilon) \subseteq \mathbb{Z}_{\max}$ for all $0 \leq |\varepsilon| \leq \varepsilon_{\max}$, it follows that for sufficiently small $|\varepsilon|$, $\mathbb{Z}(\varepsilon) \subseteq \mathcal{N}(\mathbb{Z}(0))$ for any open neighborhood $\mathcal{N}(\mathbb{Z}(0))$ of $\mathbb{Z}(0)$, and thus $\gamma(x, u; \varepsilon) > 0$ for all $(x, u) \in \mathbb{Z}(\varepsilon) \setminus B_\delta(x^*(0), u^*(0))$. Together with the above established fact that $(x^*(\varepsilon), u^*(\varepsilon))$ is a strict minimizer of γ on the set $\mathbb{Z}(\varepsilon) \cap B_\delta(x^*(0), u^*(0))$ and $\gamma(x^*(\varepsilon), u^*(\varepsilon); \varepsilon) = 0$, this implies that $(x^*(\varepsilon), u^*(\varepsilon))$ is indeed the unique global minimizer of problem $\mathcal{P}_\gamma[\varepsilon]$. However, this in particular implies that $(x^*(\varepsilon), u^*(\varepsilon))$ is also the unique global minimizer of problem $\mathcal{P}_\ell[\varepsilon]$ due to the definition of problems $\mathcal{P}_\ell[\varepsilon]$ in (4.25) and $\mathcal{P}_\gamma[\varepsilon]$ in (4.27), and the definition of γ (see (4.26)). But this means that $S^*(\varepsilon) = \{(x^*(\varepsilon), u^*(\varepsilon))\}$, i.e., $(x^*(\varepsilon), u^*(\varepsilon))$ is indeed the (unique) optimal state/input equilibrium pair.

Hence it follows that there exists $0 < \bar{\varepsilon} \leq \varepsilon_{\max}$ such that for all $|\varepsilon| \leq \bar{\varepsilon}$, the system (2.1) is strictly dissipative on $\mathbb{Z}(\varepsilon)$ with respect to the supply rate $s(x, u; \varepsilon) = \ell(x, u) - \ell(x^*(\varepsilon), u^*(\varepsilon))$, and the corresponding storage function $\lambda(x; \varepsilon)$ is defined in (4.32)–(4.33). This concludes the proof of Theorem 4.15. $\qquad \square$

Remark 4.18. *Assumption 4.1, i.e., existence of a compact set $\mathbb{Z}_{\max}$ such that $\mathbb{Z}(\varepsilon) \subseteq \mathbb{Z}_{\max}$ for all $0 \leq |\varepsilon| \leq \varepsilon_{\max}$, was needed in Part 2 of the proof of Theorem 4.15 in order to ensure*

---

[7]In case that $\mathbb{Z}_{\max} \setminus \mathcal{N}(\mathbb{Z}(0)) = \emptyset$, by convention $g_{j,\min}(\varepsilon) := \infty$.

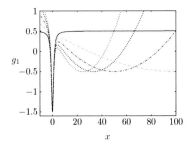

Figure 4.2: Function $g_1(x, u, \varepsilon)$ in Remark 4.18 for $\varepsilon = 0$ (solid), $\varepsilon = 0.01$ (dashed), $\varepsilon = 0.02$ (dash-dotted), $\varepsilon = 0.03$ (dotted) and $\varepsilon = 0.04$ (dotted).

that $\mathbb{Z}(\varepsilon) \subseteq \mathcal{N}(\mathbb{Z}(0))$ for any neighborhood $\mathcal{N}(\mathbb{Z}(0))$ of $\mathbb{Z}(0)$ and sufficiently small $|\varepsilon|$, which means that the set-valued map $\varepsilon \to \mathbb{Z}(\varepsilon)$ is upper semicontinuous (Aubin and Cellina, 1984) at $\varepsilon = 0$. Note that this conclusion does not hold if Assumption 4.1 is relaxed to $\mathbb{Z}(\varepsilon)$ being compact for each $0 \leq |\varepsilon| \leq \varepsilon_{\max}$. This can, e.g., be seen by considering the function $g : \mathbb{R} \times \mathbb{R} \times \mathbb{R} \to \mathbb{R}^2$ with $g_1(x, u, \varepsilon) = (x^2 - 1)/(x^2 + 1) + (\varepsilon x - 1)^2 - 1.5$ and $g_2(x, u, \varepsilon) = u^2 - 1$. Figure 4.2 shows g_1 for different values of ε. For each ε, the set $\mathbb{Z}(\varepsilon)$ defined via (4.23) is compact, but there does not exist some $\varepsilon_{\max} > 0$ and a compact set $\mathbb{Z}_{\max}$ such that $\mathbb{Z}(\varepsilon) \subseteq \mathbb{Z}_{\max}$ for all $0 \leq |\varepsilon| \leq \varepsilon_{\max}$. Namely, while for $\varepsilon = 0$ we have $\mathbb{Z}(0) = [-\sqrt{3}, \sqrt{3}] \times [-1, 1]$, x-values of arbitrary magnitude are contained in $\mathbb{Z}(\varepsilon)$ for some small enough $|\varepsilon| > 0$. This means that the set-valued map $\varepsilon \to \mathbb{Z}(\varepsilon)$ is in particular not upper semicontinuous at $\varepsilon = 0$, and hence the conclusions in Part 2 of the proof of Theorem 4.15 are not valid anymore.

In Theorem 4.15, robustness of dissipativity with respect to small changes in the constraint set $\mathbb{Z}$ was established. These results can be extended in various ways, as discussed in the following remark.

Remark 4.19. *We note that the results of Theorem 4.15 can be extended in a straightforward way to the case where also the stage cost function ℓ and the system dynamics f depend on the additional parameters ε, i.e., robustness of dissipativity with respect to small changes in the cost function and the system dynamics can be established. Furthermore, the results of Theorem 4.15 can also be extended to an economic MPC setting including average constraints as introduced in Section 2.2.2, where dissipativity with respect to the (relaxed) supply rate $\bar{s}(x, u) = \ell(x, u) - \ell(x^*, u^*) + \bar{\lambda}^T h(x, u)$ for some (arbitrary) $\bar{\lambda} \in \mathbb{R}^p_{\geq 0}$ is of interest. Namely, similar to the proof of Theorem 4.15, one can show that both the storage function λ and the multiplier $\bar{\lambda}$ can be modified continuously in ε such that strict dissipativity is maintained under small changes in ε. Finally, Theorem 4.15 can also be extended to the case of general parameter dependent supply rates $s(x, u; \varepsilon)$ different from the specific one used in economic MPC, which means that the obtained results are also of interest beyond economic MPC, see (Müller et al., 2014d).*

Convex case

We now show that further results beyond the robustness analysis of Theorem 4.15 can be obtained if a certain convexity assumption is satisfied. Namely, instead of considering small perturbations in the constraint set as expressed by the parameter ε, we now look at additional constraints g_{ad} which are imposed on the system and can possibly alter the optimal steady-state to a large extend. Hence in the following, with a slight change of notation, we drop the dependence of the various functions and optimization problems on ε, but instead define the sets $\mathbb{Z} := \{(x, u) : g(x, u) \leq 0\}$, $\mathbb{Z}_{ad} := \{(x, u) : g(x, u) \leq 0, g_{ad}(x, u) \leq 0\}$, $S_{ad} := \{(x, u) \in \mathbb{Z}_{ad} : x = f(x, u)\}$ and $S_{ad}^* := \{(y, w) \in S_{ad} : \ell(y, w) = \min\limits_{(x,u) \in S_{ad}} \ell(x, u)\}$. As above, let (x_{ad}^*, u_{ad}^*) denote an arbitrary element of S_{ad}^*, and define the optimization problems $\mathcal{P}_{\ell, ad} := \mathcal{P}([x^T\ u^T]^T, \ell, x - f(x, u), [g^T\ g_{ad}^T]^T)$ and $\mathcal{P}_{\gamma ad} := \mathcal{P}([x^T\ u^T]^T, \gamma_{ad}, 0, [g^T\ g_{ad}^T]^T)$ as in (B.1) with

$$\gamma_{ad}(x, u) := \ell(x, u) - \ell(x_{ad}^*, u_{ad}^*) + \lambda_{ad}(x) - \lambda_{ad}(f(x, u)) \qquad (4.35)$$

for some storage function $\lambda_{ad}(x)$. The following Theorem shows that if γ as defined in (4.21) is convex, then for each feasible steady-state $(y, w) \in S$ there exists a function g_{ad} such that (y, w) is an optimal steady-state under the additional constraints, i.e., $(y, w) \in S_{ad}^*$, and the dissipativity property with respect to the new supply rate $s_{ad}(x, u) = \ell(x, u) - \ell(x_{ad}^*, u_{ad}^*)$ is maintained with the same storage function $\lambda_{ad}(x) = \lambda(x)$.

Theorem 4.20. *Suppose that the following in satisfied:*

(i) The functions f, g and ℓ are continuously differentiable in (x, u) and g is convex.

(ii) There exist a continuously differentiable storage function λ such that system (2.1) without additional constraints is (strictly) dissipative on $\mathbb{Z}$ with respect to the supply rate $s(x, u) = \ell(x, u) - \ell(x^*, u^*)$, and the corresponding function γ defined in (4.21) is (strictly) convex on $\mathbb{Z}$.

Then, for each feasible state/input equilibrium pair $(y, w) \in S$, there exists an additional constraint function g_{ad} which is convex and continuously differentiable in (x, u) such that $(y, w) \in S_{ad}^*$ and the system (2.1) is (strictly) dissipative on $\mathbb{Z}_{ad}$ with respect to the supply rate $s_{ad}(x, u) = \ell(x, u) - \ell(x_{ad}^*, u_{ad}^*)$ and with storage function $\lambda_{ad}(x) = \lambda(x)$.

Conversely, if for a given convex and continuously differentiable additional constraint function g_{ad}, a state/input equilibrium pair $(y, w) \in S$ together with some $\nu = \begin{bmatrix} \nu_g^T & \nu_{g_{ad}}^T \end{bmatrix}^T$ satisfies the KKT conditions (B.2)–(B.3) for problem $\mathcal{P}_{\gamma, ad}$ with $\lambda_{ad}(x) = \lambda(x)$, then $(y, w) \in S_{ad}^*$ and system (2.1) is (strictly) dissipative on $\mathbb{Z}_{ad}$ with respect to the supply rate $s_{ad}(x, u) = \ell(x, u) - \ell(x_{ad}^*, u_{ad}^*)$ and with storage function $\lambda_{ad}(x) = \lambda(x)$.

Proof. As γ is (strictly) convex by assumption, also γ_{ad} defined in (4.35) with $\lambda_{ad}(x) := \lambda(x)$ is (strictly) convex since the two functions only differ by a constant term. Hence, in order to establish that a given steady-state $(y, w) \in S$ is a global minimizer (the unique global minimizer) of problem $\mathcal{P}_{\gamma ad}$ with $\lambda_{ad}(x) = \lambda(x)$, by Proposition B.4 it is sufficient to show that there exists a function $g_{ad}(x, u)$ which is convex and continuously differentiable in (x, u), such that the KKT conditions (B.2)–(B.3) are satisfied at (y, w). It is easily seen that for each steady-state $(y, w) \in S$, this is possible by choosing, e.g., g_{ad} as a scalar linear function $g_{ad}(x, u) = [x^T\ u^T]a + b$, where $a \in \mathbb{R}^{n+m}$ and $b \in \mathbb{R}$ are such that

(i) $a = -(\nabla_{(x,u)}\gamma(y,w))^T$ if $\nabla_{(x,u)}\gamma(y,w) \neq 0$ ($a \neq 0$ otherwise) and (ii) $g_{ad}(y,w) = 0$. Then, (B.2)–(B.3) are satisfied with $\nu_{g_{ad}} = 1$ and $\nu_g = 0$ if $\nabla_{(x,u)}\gamma(y,w) \neq 0$, and $\nu_{g_{ad}} = 0$ and $\nu_g = 0$ otherwise. Furthermore, due to the definition of problems $\mathcal{P}_{\ell,ad}$ and $\mathcal{P}_{\gamma_{ad}}$ and the definition of γ_{ad}, each global minimizer of problem $\mathcal{P}_{\gamma_{ad}}$ which is a steady-state is also a global minimizer of problem $\mathcal{P}_{\ell,ad}$, i.e., $(y,w) \in S^*_{ad}$. In case that γ is convex, this implies that the system (2.1) is dissipative on $\mathbb{Z}_{ad}$ with respect to the supply rate $s_{ad}(x,u) = \ell(x,u) - \ell(x^*_{ad}, u^*_{ad})$ and with storage function $\lambda_{ad}(x) = \lambda(x)$. In case that γ is strictly convex, (y,w) being the unique global minimizer of problem $\mathcal{P}_{\gamma_{ad}}$ implies that there exists a function $\alpha' \in \mathcal{K}_\infty$ such that $\gamma_{ad}(x,u) \geq \alpha'(|(x-y, u-w)|)$ for all $(x,u) \in \mathbb{Z}_{ad}$. Namely, in case that $\mathbb{Z}_{ad}$ is compact, this follows as discussed in Remark 4.17; in case that $\mathbb{Z}_{ad}$ is unbounded, by strict convexity of γ_{ad} it follows that γ_{ad} is radially unbounded, from where again existence of a function α' satisfying the above can be concluded (see, e.g., Damm et al. (2014)). Hence the system (2.1) is strictly dissipative on $\mathbb{Z}_{ad}$ with respect to the supply rate $s_{ad}(x,u) = \ell(x,u) - \ell(x^*_{ad}, u^*_{ad})$ and with storage function $\lambda_{ad}(x) = \lambda(x)$.

The second statement of the Theorem directly follows from Proposition B.4 and the above considerations that each global minimizer of problem $\mathcal{P}_{\gamma_{ad}}$ which is a steady-state is also a global minimum of problem $\mathcal{P}_{\ell,ad}$, and that (y,w) being a global minimizer (the unique global minimizer) of problem $\mathcal{P}_{\gamma_{ad}}$ implies that the system (2.1) is dissipative (strictly dissipative) on $\mathbb{Z}_{ad}$. $\qquad\square$

Remark 4.21. *Note that in contrast to Theorem 4.15, where compactness of the constraint sets $\mathbb{Z}(\varepsilon)$ was required (compare also Remark 4.18), the results of Theorem 4.20 are also valid for a possibly unbounded constraint set $\mathbb{Z}$. Furthermore, the convexity assumption on γ in condition (ii) of Theorem 4.20 is always satisfied if ℓ is strictly convex, system (2.1) is linear and Slater's condition is satisfied, i.e., there exists some (x,u) with $g(x,u) < 0$. Namely, in this case, due to strong duality the storage function λ in (4.21) can be chosen as a linear function[8] (Damm et al., 2014; Diehl et al., 2011). In (Damm et al., 2014), also some extensions to the case where ℓ is convex (but not strictly convex) are presented, in which case the storage function λ can be chosen as a quadratic function under certain conditions. However, note that for γ to be convex, neither ℓ has to be convex nor the system (2.1) has to be linear. For instance, in Example 4.13, the cost function ℓ was convex but the system was nonlinear, yet the function γ was still convex.*

The convexity assumption on γ discussed above was the crucial assumption in Theorem 4.20 which allowed us to establish robustness of dissipativity with respect to possibly large changes in the constraints, compared to only small ones as in Theorem 4.15. On the other hand, if γ is not convex, one can show that the statements of Theorem 4.20 still hold for each steady-state $(y,w) \in S$ such that γ can be lower bounded on the convex set $\mathbb{Z}$ by a (strictly) convex and continuously differentiable function $\hat{\gamma}$ satisfying $\hat{\gamma}(y,w) = \gamma(y,w)$. Namely, as $\hat{\gamma}$ is (strictly) convex, one can establish as in the proof of Theorem 4.20 that (y,w) (uniquely) minimizes $\hat{\gamma}$ over $\mathbb{Z}_{ad}$. But then, as $\hat{\gamma}$ is a lower bound for γ on $\mathbb{Z}$ and furthermore $\hat{\gamma}(y,w) = \gamma(y,w)$, it follows that (y,w) also (uniquely) minimizes γ, and hence also γ_{ad}, over $\mathbb{Z}_{ad}$, i.e., the system is again (strictly) dissipative on $\mathbb{Z}_{ad}$ with respect to the supply rate $s_{ad}(x,u) = \ell(x,u) - \ell(x^*_{ad}, u^*_{ad})$ and with storage function $\lambda_{ad}(x) = \lambda(x)$.

---

[8]Boundedness from below of the storage function λ as required in Definition 2.7 is then given if the set $\mathbb{Z}$ is compact.

Example 4.22. *We now revisit our motivating Example 4.13 and explain why the dissipativity property was lost. Namely, for $\varepsilon = 0$, $(x^*(0), u^*(0)) = (1, 1)$ is the unique global minimizer of problems $\mathcal{P}_\ell[0]$ and $\mathcal{P}_\gamma[0]$, and also the strong second order sufficiency conditions are satisfied. The only assumption of Theorem 4.15 which is not satisfied is the regularity assumption on $(x^*(0), u^*(0))$. Namely, for problem $\mathcal{P}_\ell[0]$, $\nabla_{(x,u)} h_s(1, 1)$ and $\nabla_{(x,u)} g_{ad}(1, 1; 0)$ are not linearly independent. This is the reason why for arbitrarily small $\varepsilon > 0$, the optimal feasible steady-state "jumps", and hence dissipativity with respect to the supply rate $s(x, u; \varepsilon)$ is lost. On the other hand, when, e.g., an additional constraint of the form $g_{ad}(x, u; \varepsilon) = x + u - 2 + \varepsilon \leq 0$ is considered, for small $\varepsilon > 0$ we obtain $(x^*(\varepsilon), u^*(\varepsilon)) = (1 - \varepsilon, 1)$, i.e., the optimal steady-state changes continuously with ε. It is straightforward to verify that all the assumptions of Theorem 4.15 are satisfied[9], and hence the system remains dissipative for small ε. Furthermore, as γ given by (4.21) is convex, we can apply the results of Theorem 4.20. For example, if the additional constraint $\hat{g}_{ad}(x, u) = (1 - 3\delta(1 - \bar{u}))x + \delta(2 + \bar{u})u + 2\delta(2 + \bar{u}) \leq 0$ is applied (see Figure 4.1(b)), then $S^*_{ad} = \{(0, -2)\}$ and by Theorem 4.20 we conclude that the system is still dissipative with respect to the new supply rate $s_{ad}(x, u) = \ell(x, u) - \ell(0, -2) = \ell(x, u) - (1 + \delta(2 + \bar{u})^2)$ and with the same storage function $\lambda_{ad}(x) = \lambda(x) = 2\delta(1 - \bar{u})x$.*

Summary

In this section, we have examined robustness of dissipativity in economic MPC. In particular, we have shown that under certain regularity assumptions, system (2.1) is robustly dissipative with respect to small changes in the constraint set $\mathbb{Z}$. Furthermore, in case that the involved function γ is convex, the system also remains dissipative for certain possibly large changes in the constraint set. We note that these results are also of interest in the context of distributed cooperative control of multiple systems as in Chapters 3 and 5, for example in order to determine whether the overall system (including coupling constraints) is dissipative given that each single system is dissipative.

4.2 Average constraints

We now turn our attention to economic MPC including average constraints, as introduced in Section 2.2.2. As already stated there, the contributions of this section are (i) to develop an economic MPC framework which guarantees satisfaction of asymptotic average constraints in a terminal region setting (see Section 4.2.1); (ii) to show that the closed-loop system asymptotically converges to the optimal steady-state x^* under an appropriate dissipativity condition (see Section 4.2.2); (iii) to develop an economic MPC framework including *transient* in contrast to asymptotic average constraints (see Section 4.2.3).

---

[9]Here, for $\varepsilon = 0$ strict complementarity does not hold (as both $g_{ad}(1, 1; 0)$ as well as the corresponding Lagrange multiplier are zero). Hence, as pointed out in Remark B.3 in Appendix B, the strong second order sufficiency condition as in Assumption (ii) of Theorem 4.15 is needed.

4.2.1 Economic MPC with average constraints and terminal region

Consider again the economic MPC setting of Section 2.2.2 with asymptotic average constraints, namely system (2.1) subject to the pointwise-in-time state and input constraints (2.2) and the additional asymptotic average constraints (2.14), i.e., $(x(t), u(t)) \in \mathbb{Z} \subseteq \mathbb{X} \times \mathbb{U}$ for all $t \in \mathbb{I}_{\geq 0}$ and $Av[h(\boldsymbol{x}, \boldsymbol{u})] \subseteq \mathbb{Y}$. In the following, by an optimal state/input equilibrium pair (x^*, u^*) we denote an arbitrary element of the set S_{av}^* as defined in (2.16). In order to ensure that the asymptotic average constraints are satisfied for the closed-loop system, we have to modify Problem 2.2 as follows. Namely, at each time $t \in \mathbb{I}_{\geq 0}$ with measured state $x(t)$, the following optimization problem is solved:

Problem 4.23.

$$\underset{\boldsymbol{u}(t)}{minimize} \; J_N(x(t), \boldsymbol{u}(t))$$

subject to

$$x(k+1|t) = f(x(k|t), u(k|t)), \quad k \in \mathbb{I}_{[0,N-1]}, \tag{4.36a}$$

$$x(0|t) = x(t), \tag{4.36b}$$

$$(x(k|t), u(k|t)) \in \mathbb{Z}, \quad k \in \mathbb{I}_{[0,N-1]}, \tag{4.36c}$$

$$x(N|t) \in \mathbb{X}^f(t), \tag{4.36d}$$

$$\sum_{k=0}^{N-1} h(x(k|t), u(k|t)) \in \mathbb{Y}_t, \tag{4.36e}$$

where

$$J_N(x(t), \boldsymbol{u}(t)) := \sum_{k=0}^{N-1} \ell(x(k|t), u(k|t)) + V^f(x(N|t)). \tag{4.37}$$

In this problem, $\boldsymbol{u}(t) := \{u(0|t), \ldots, u(N-1|t)\}$ and $\boldsymbol{x}(t) := \{x(0|t), \ldots, x(N|t)\}$ are again input and corresponding state sequences predicted at time t over the prediction horizon $N \in \mathbb{I}_{\geq 0}$. Compared to Problem 2.2, the differences of Problem 4.23 are that (i) it includes the additional constraint (4.36e), which is needed in order to ensure satisfaction of the asymptotic average constraints (2.14), and (ii) we need to allow for a possibly time-varying terminal region $\mathbb{X}^f(t)$ in (4.36d) for reasons to become apparent in the following. The time-varying output set $\mathbb{Y}_t$ in (4.36e) is recursively defined as

$$\mathbb{Y}_{t+1} := \mathbb{Y}_t \oplus \mathbb{Y} \oplus \overline{\mathbb{Y}}(t) \oplus \{-h(x(t), u(t))\}, \quad \mathbb{Y}_0 := N\mathbb{Y} \oplus \mathbb{Y}_{00}, \tag{4.38}$$

where $\mathbb{Y}_{00} \subseteq \mathbb{R}^p$ is an arbitrary compact set and $\overline{\mathbb{Y}}(t)$ will be specified later. We remark that the recursion in (4.38) can be solved explicitly, yielding (thanks to convexity of $\mathbb{Y}$)

$$\mathbb{Y}_t = \mathbb{Y}_{00} \oplus (t+N)\mathbb{Y} \oplus \sum_{k=0}^{t-1} \overline{\mathbb{Y}}(k) \oplus \left\{ \sum_{k=0}^{t-1} -h(x(k), u(k)) \right\}. \tag{4.39}$$

Remark 4.24. *As already mentioned in Section 2.2.2, an economic MPC scheme similar to Problem 4.23 for a setting including asymptotic average constraints was initially proposed in (Angeli et al., 2012, Section V.B), compare also (Angeli et al., 2011). The main novel features of the scheme proposed in this section are that firstly we adopt the more general terminal region and terminal cost approach, while a terminal equality constraint $x(N|t) = x^*$ was used in the above references. Secondly, we use a more general definition of the time-varying output constraint set $\mathbb{Y}_t$ in (4.38), including the set $\overline{\mathbb{Y}}(t)$. In a terminal equality constraint setting, this relaxation can be used for (transient) performance improvement, see (Angeli et al., 2011), where the special choice $\overline{\mathbb{Y}}(t) = ((t+1)^\alpha - t^\alpha)\overline{\mathbb{Y}}$ for some compact set $\overline{\mathbb{Y}}$ and some $0 \leq \alpha < 1$ was used. On the other hand, within the adopted terminal region and terminal cost approach, this more general definition of the time-varying output constraint set $\mathbb{Y}_t$ in (4.38) is in fact necessary in order to guarantee recursive feasibility and fulfillment of the average constraints, as shown below in more detail.*

Remark 4.25. *As noted above, the set $\mathbb{Y}_{00} \subseteq \mathbb{R}^p$ in (4.38) is an arbitrary compact set and can be used as a tuning parameter (compare Angeli et al. (2011)). Namely, a larger set $\mathbb{Y}_{00}$ implies that the constraint (4.36e) is (initially) less restrictive, and hence the system has more freedom to "spend time" in a region where $h(x, u) \notin \mathbb{Y}$. This typically results in an improved (transient) performance of the closed-loop system (compare the examples in Sections 4.2.2, 4.2.4 and 5.3). In any case, $\mathbb{Y}_{00}$ has to be large enough such that constraint (4.36e) is feasible at time $t = 0$.*

As already mentioned above, the terminal region $\mathbb{X}^f(t)$ in (4.36d) is possibly time-varying. We then also need to slightly adapt Assumption 2.5 as follows.

Assumption 4.3. *For each $t \in \mathbb{I}_{\geq 0}$, the terminal region $\mathbb{X}^f(t) \subseteq \mathbb{X}$ is closed and $x^* \in \mathbb{X}^f(t)$. Furthermore, there exists a local auxiliary control law $u = \kappa^f(x)$ such that the following is satisfied for all $t \in \mathbb{I}_{\geq 0}$ and all $x \in \mathbb{X}^f(t)$:*

i) $(x, \kappa^f(x)) \in \mathbb{Z}$

ii) $f(x, \kappa^f(x)) \in \mathbb{X}^f(t+1)$

iii) $V^f(f(x, \kappa^f(x))) - V^f(x) \leq -\ell(x, \kappa^f(x)) + \ell(x^*, u^*)$

Remark 4.26. *In case that the terminal region is constant, i.e., $\mathbb{X}^f(t) = \mathbb{X}^f$ for all $t \in \mathbb{I}_{\geq 0}$, condition ii) of Assumption 4.3 reduces to Assumption 2.5ii). Furthermore, note that while in Assumption 2.5, we required x^* to lie in the interior of $\mathbb{X}^f$ (which was needed in order to ensure asymptotic stability), we now only require that it is contained in $\mathbb{X}^f(t)$ for all $t \in \mathbb{I}_{\geq 0}$ (but not necessarily in the interior). This is the case as in economic MPC including average constraints, one can, under a suitable dissipativity condition, only ensure asymptotic convergence of the closed-loop system but not asymptotic stability (see Section 4.2.2), and hence the assumption that x^* lies in the interior of the terminal region is not needed.*

Besides Assumption 4.3, in order to ensure satisfaction of the asymptotic average constraints (2.14), we furthermore need the following conditions concerning the sets $\overline{\mathbb{Y}}(t)$.

Assumption 4.4. *For each $t \in \mathbb{I}_{\geq 0}$, the set $\overline{\mathbb{Y}}(t)$ is such that $h(x, \kappa^f(x)) \in \mathbb{Y} \oplus \overline{\mathbb{Y}}(t)$ for all $x \in \mathbb{X}^f(t)$.*

Assumption 4.5. *There exists a sequence* $\boldsymbol{\sigma} : \mathbb{I}_{\geq 0} \to \mathbb{R}_{\geq 0}$ *with* $\lim_{t \to \infty} \sigma(t)/t = 0$ *such that* $\sum_{k=0}^{t-1} \overline{\mathbb{Y}}(k) \subseteq \sigma(t) B_1(0)$ *for all* $t \in \mathbb{I}_{\geq 1}$.

Assumptions 4.4 and 4.5 will be crucial in establishing recursive feasibility of problem 4.23 and fulfillment of the average constraints. We will show below how the sets $\mathbb{X}^f(t)$ and $\overline{\mathbb{Y}}(t)$ can be calculated such that Assumptions 4.3–4.5 are satisfied.

Remark 4.27. *Assumption 4.5 is slightly weaker than Assumption 3 in the paper of Müller et al. (2014f), on which the results of this section are based. Namely, there it was assumed that* $\sum_{k=0}^{t-1} \overline{\mathbb{Y}}(k) \subseteq t^\alpha \overline{\mathbb{Y}}$ *for all* $t \in \mathbb{I}_{\geq 0}$, *some constant* $0 \leq \alpha < 1$ *and some compact set* $\overline{\mathbb{Y}}$, *which means that Assumption 4.5 holds with* $\sigma(t) = ct^\alpha$ *for some* $c > 0$ *and* $0 \leq \alpha < 1$.

Now as in Chapter 2, denote again the minimizer of Problem 4.23 by $\boldsymbol{u}^0(t)$, the corresponding state sequence by $\boldsymbol{x}^0(t)$, and by $\mathbb{X}_N(t)$ the set of all states $x \in \mathbb{X}$ such that Problem 4.23 (with $x(t) = x$) has a solution. Note that $\mathbb{X}_N(t)$ is possibly time-varying, which results from the fact that both the terminal region $\mathbb{X}^f(t)$ and the output constraint set $\mathbb{Y}_t$ are time-varying. Nevertheless, in Theorem 4.29 (see below) it is shown that recursive feasibility of Problem 4.23 follows from initial feasibility, which means that the feasible set of the following algorithm is, in fact, $\mathbb{X}_N(0)$. The economic model predictive control scheme including average constraints is now defined as follows.

Algorithm 4.28 (Economic MPC with asymptotic average constraints). *Consider system* (2.1). *At each time* $t \in \mathbb{I}_{\geq 0}$, *measure the state* $x(t)$, *solve Problem 4.23 and apply the control input* $u(t) := u^0(0|t)$.

In the following, we show that the properties of Algorithm 4.28 are as desired and that in particular satisfaction of the asymptotic average constraints (2.14) is guaranteed for the resulting closed-loop system.

Theorem 4.29. *Let* $x_0 \in \mathbb{X}_N(0)$, *and suppose that Assumptions 2.1, 2.3, 2.6, 2.7, and 4.3–4.5 are satisfied. Then Problem 4.23 is feasible for all* $t \in \mathbb{I}_{\geq 0}$ *and the following is satisfied for the closed-loop system* (2.6) *resulting from application of Algorithm 4.28:*

$$(x(t), u(t)) \in \mathbb{Z}, \quad t \in \mathbb{I}_{\geq 0}, \tag{4.40}$$

$$Av[h(\boldsymbol{x}, \boldsymbol{u})] \subseteq \mathbb{Y}, \tag{4.41}$$

$$\limsup_{T \to \infty} \frac{\sum_{t=0}^{T} \ell(x(t), u(t))}{T+1} \leq \ell(x^*, u^*). \tag{4.42}$$

Proof. As discussed in Remark 4.24, the economic MPC scheme proposed in this section is a generalization of the one in (Angeli et al., 2012, Section V.B) and (Angeli et al., 2011); hence also the proof of Theorem 4.29 follows and extends the proof of (Angeli et al., 2012, Theorem 5) and (Angeli et al., 2011, Proposition 2.3), respectively. Now suppose that Problem 4.23 is feasible at time $t \in \mathbb{I}_{\geq 0}$. Then, at time $t + 1$, consider the usual candidate input sequence $\tilde{\boldsymbol{u}}(t + 1) := \{u^0(1|t), \ldots, u^0(N - 1|t), \kappa^f(x^0(N|t))\}$ with corresponding candidate state sequence $\tilde{\boldsymbol{x}}(t+1) := \{x^0(1|t), \ldots, x^0(N|t), f(x^0(N|t), \kappa^f(x^0(N|t)))\}$. These candidate sequences fulfill the constraints (4.36a)–(4.36d) due to Assumption 4.3 and the fact that the minimizer $\boldsymbol{u}^0(t)$ of Problem 4.23 at time t was feasible. Furthermore, due to

Assumption 4.4 and the fact that $\boldsymbol{u}^0(t)$ and $\boldsymbol{x}^0(t)$ satisfy (4.36e), we obtain

$$\sum_{k=0}^{N-1} h(\tilde{x}(k|t+1), \tilde{u}(k|t+1))$$

$$= \sum_{k=0}^{N-1} h(x^0(k|t), u^0(k|t)) + h(x^0(N|t), \kappa^f(x^0(N|t))) - h(x(t), u(t))$$

$$\in \mathbb{Y}_t \oplus \mathbb{Y} \oplus \overline{\mathbb{Y}}(t) \oplus \{-h(x(t), u(t))\} = \mathbb{Y}_{t+1}.$$

Hence Problem 4.23 is feasible for all $t \in \mathbb{I}_{\geq 0}$, and from the definition the input $u(t) := u^0(0|t)$ in Algorithm 4.28, it immediately follows that the resulting closed-loop system (2.6) satisfies the pointwise-in-time state and input constraints (4.40).

Next, in order to show that the closed-loop system (2.6) satisfies the average constraints (4.41), consider the following. From (4.36e) and (4.39), it follows that for all $t \in \mathbb{I}_{\geq 0}$

$$\sum_{k=0}^{t-1} h(x(k), u(k)) + \sum_{k=0}^{N-1} h(x(k|t), u(k|t)) \in \mathbb{Y}_{00} \oplus (t+N)\mathbb{Y} \oplus \sum_{k=0}^{t-1} \overline{\mathbb{Y}}(k), \qquad (4.43)$$

for each predicted input and state sequences which are feasible at time t. By Assumptions 2.6 and 2.7, i.e., compactness of $\mathbb{Z}$ and continuity of h, for any feasible predicted input and state sequences we have

$$\lim_{t \to +\infty} \frac{\sum_{k=0}^{N-1} h(x(k|t), u(k|t))}{t} = 0. \qquad (4.44)$$

Hence for any infinite subsequence $\{t_n\} \subseteq \mathbb{I}_{\geq 0}$ such that $\lim_{n \to +\infty} (\sum_{k=0}^{t_n-1} h(x(k), u(k))/t_n)$ exists, it follows that

$$\lim_{n \to +\infty} \frac{\sum_{k=0}^{t_n-1} h(x(k), u(k))}{t_n} \stackrel{(4.43),(4.44)}{\in} \lim_{n \to +\infty} \frac{\mathbb{Y}_{00} \oplus (t_n + N)\mathbb{Y} \oplus \sum_{k=0}^{t_n-1} \overline{\mathbb{Y}}(k)}{t_n}$$

$$\stackrel{\text{Ass. 4.5}}{\subseteq} \lim_{n \to +\infty} \frac{\mathbb{Y}_{00} \oplus (t_n + N)\mathbb{Y} \oplus \sigma(t_n)B_1(0)}{t_n} = \mathbb{Y}. \qquad (4.45)$$

Here, the set-limit has to be understood as in (Goebel et al., 2012, Definition 5.1), and the last equality holds due to compactness of $\mathbb{Y}_{00}$ and the fact that $\lim_{t \to \infty} \sigma(t)/t = 0$ according to Assumption 4.5. Hence (4.41) is satisfied.

Finally, the average performance estimate (4.42) follows as in case without average constraints, i.e., as in Theorem 2.5. $\qquad \Box$

Remark 4.30. *In Theorem 4.29, relaxing Assumption 2.6, i.e., allowing for an unbounded constraint set $\mathbb{Z}$, can be done as discussed in the paragraph below Theorem 2.5. However, one then in addition needs to assume that h is bounded on $\mathbb{Z}$ in order to ensure that $Av[h(\boldsymbol{x}, \boldsymbol{u})]$ is well defined and (4.44) is satisfied for any feasible predicted input and state sequences, which in the proof of Theorem 4.29 followed from compactness of $\mathbb{Z}$ and continuity of h.*

Calculating $\mathbb{X}^f(t)$ and $\overline{\mathbb{Y}}(t)$

We now discuss how the sets $\mathbb{X}^f(t)$ and $\overline{\mathbb{Y}}(t)$ can be calculated such that Assumptions 4.3–4.5 are satisfied. First, we remark that in the case of a terminal equality constraint $x(N|t) = x^*$ (where Assumption 4.3 is trivially satisfied with the constant terminal region $\mathbb{X}^f = \{x^*\}$ and $\kappa^f(x^*) = u^*$), one can choose $\overline{\mathbb{Y}}(t) = \{0\}$ for all $t \in \mathbb{I}_{\geq 0}$ and Assumptions 4.4 and 4.5 are immediately satisfied (with $\sigma \equiv 0$), i.e., the results of (Angeli et al., 2012, Section V.B) are recovered. In this case, as also pointed out in Remark 4.24, choosing $\overline{\mathbb{Y}}(t) \supsetneq \{0\}$ can be used to relax the output constraint (4.36e) in order to improve the transient performance of the system. Loosely speaking, this means that the system is allowed to spend some more time in a region where $h(x, u) \notin \mathbb{Y}$.

We now turn our attention to the setting with a terminal region constraint as in (4.36d) instead of a terminal equality constraint. If a (nontrivial) constant terminal region $\mathbb{X}^f$ can be found such that $h(x, \kappa^f(x)) \in \mathbb{Y}$ for all $x \in \mathbb{X}^f$, then the same considerations as above apply, i.e., $\overline{\mathbb{Y}}(t) = \{0\}$ for all $t \in \mathbb{I}_{\geq 0}$ is a possible choice such that Assumptions 4.4 and 4.5 are satisfied with $\sigma \equiv 0$; note that in such a case, there is also no need for a time-varying terminal region. On the other hand, if no (nontrivial) constant terminal region $\mathbb{X}^f$ can be found such that $h(x, \kappa^f(x)) \in \mathbb{Y}$ for all $x \in \mathbb{X}^f$, things are inherently different as Assumption 4.4 is not satisfied anymore with $\overline{\mathbb{Y}}(t) = \{0\}$. Hence in this case, a nontrivial $\overline{\mathbb{Y}}(t)$ is not only beneficial in order to improve performance, but is first of all needed in order to establish recursive feasibility of Problem 4.23 via Theorem 4.29. In the following, we will construct sets $\mathbb{X}^f(t)$ and $\overline{\mathbb{Y}}(t)$ satisfying Assumptions 4.3–4.5; this will be done on the basis of a constant terminal region $\mathbb{X}^f$ satisfying Assumption 2.5 (which, as discussed in Section 2.2, can, e.g., be calculated as proposed by Amrit et al. (2011)). Furthermore, the following assumption is imposed on the constant terminal region $\mathbb{X}^f$ and the local auxiliary control law κ^f:

Assumption 4.6. *There exists a (constant) terminal region $\mathbb{X}^f$ satisfying Assumption 2.5, which is of the form $\mathbb{X}^f := \{x \in \mathbb{R}^n : E(x) \leq \alpha_0\}$ for some $\alpha_0 > 0$ and $E(x) := (x - x^*)^T P (x - x^*)$ with $P \succ 0$. Furthermore, the following is satisfied: i) $E(f(x, \kappa^f(x))) - E(x) \leq -(x - x^*)^T Q (x - x^*)$ for some $Q \succ 0$ and all $x \in \mathbb{X}^f$, and ii) $h(x^*, \kappa^f(x^*)) \in \mathbb{Y}$.*

Remark 4.31. *Assumption 4.6i) slightly strengthens Assumption 2.5ii) from invariance of the terminal region to contractiveness, if the local auxiliary control law is applied. Note that it is sufficient if Assumption 4.6i) is satisfied for some arbitrary positive definite Q. Furthermore, as $E(x) > 0$ for all $x \neq x^*$, it follows from Assumption 4.6i) that $\kappa^f(x^*) = \bar{u}$ for some $\bar{u} \in \mathbb{U}$ satisfying $f(x^*, \bar{u}) = x^*$. This has the following two consequences. First, Assumption 4.3iii) implies in this case that $\ell(x^*, \bar{u}) \leq \ell(x^*, u^*)$, which together with $h(x^*, \bar{u}) \in \mathbb{Y}$ as ensured by Assumption 4.6ii) in fact implies that $(x^*, \bar{u}) \in S_{av}^*$ (compare (2.16)), i.e., $\kappa^f(x^*) = u^*$ without loss of generality. Second, we have that condition ii) of Assumption 4.6 is implied by condition i) in case that $h(x^*, \bar{u}) \in \mathbb{Y}$ for all $\bar{u} \in \mathbb{U}$ satisfying $f(x^*, \bar{u}) = x^*$ (which in particular is the case if u^* is the unique steady-state input corresponding to the steady-state x^*). Finally, we note that, e.g., the approach proposed by Amrit et al. (2011) results in a (constant) terminal region $\mathbb{X}^f$ and a local auxiliary control law κ^f which are such that Assumption 4.6 is satisfied (with $\kappa^f(x^*) = u^*$).*

Now suppose that a constant terminal region $\mathbb{X}^f$ is given as specified in Assumption 4.6, and define $\alpha_{\max} := \sup\{\alpha \,|\, h(x, \kappa_f(x)) \in \mathbb{Y} \,\forall x \in \mathbb{R}^n \text{ s.t. } E(x) \leq \alpha\}$. Note that $\alpha_{\max} \geq 0$

due to the fact that $h(x^*, \kappa^f(x^*)) \in \mathbb{Y}$ as required by Assumption 4.6ii). As noted above, for $\alpha_0 \leq \alpha_{\max}$, $\overline{\mathbb{Y}}(t) = \{0\}$ for all $t \in \mathbb{I}_{\geq 0}$ is a possible choice such that Assumptions 4.4 and 4.5 are satisfied (together with the constant terminal region $\mathbb{X}^f$). We now propose to gradually tighten the terminal region such that $\mathbb{X}^f(t) \subseteq \{x \in \mathbb{R}^n : E(x) \leq \alpha_{\max}\}$ at least asymptotically. Namely, for each $t \in \mathbb{I}_{\geq 0}$, let

$$\mathbb{X}^f(t) := \{x \in \mathbb{R}^n : E(x) \leq \alpha(t)\}, \tag{4.46}$$

where

$$\alpha(t+1) = \max\left\{\left(1 - \frac{\lambda_{\min}(Q)}{\lambda_{\max}(P)}\right)\alpha(t), \min\{\alpha_0, \alpha_{\max}\}\right\}, \qquad \alpha(0) = \alpha_0. \tag{4.47}$$

Proposition 4.32. *Suppose that a constant terminal region $\mathbb{X}^f$ is known satisfying Assumption 4.6. Then Assumption 4.3 is satisfied for the time-varying terminal regions $\mathbb{X}^f(t)$ given by* (4.46)–(4.47).

Proof. Fulfillment of items i) and iii) of Assumption 4.3 is inherited from the fact that these conditions were satisfied for the constant terminal region $\mathbb{X}^f$ according to Assumption 2.5i) and iii). In order to verify item ii), consider the following. As $\mathbb{X}^f(t) \subseteq \mathbb{X}^f(0) = \mathbb{X}^f$ for all $t \in \mathbb{I}_{\geq 0}$, by Assumption 4.6 we obtain that for all $x \in \mathbb{X}^f(t)$,

$$E(f(x, \kappa^f(x))) \leq E(x) - (x - x^*)^T Q (x - x^*) \leq E(x) - \lambda_{\min}(Q)|x - x^*|^2$$
$$\leq E(x) - \frac{\lambda_{\min}(Q)}{\lambda_{\max}(P)}E(x) = \left(1 - \frac{\lambda_{\min}(Q)}{\lambda_{\max}(P)}\right)E(x) \leq \left(1 - \frac{\lambda_{\min}(Q)}{\lambda_{\max}(P)}\right)\alpha(t). \tag{4.48}$$

But this means according to (4.46)–(4.47) that $f(x, \kappa^f(x)) \in \mathbb{X}^f(t+1)$ for all $x \in \mathbb{X}^f(t)$, i.e., item ii) of Assumption 4.3 is satisfied. $\square$

Remark 4.33. *We note that the use of time-varying terminal regions* (4.46) *within Problem 4.23 leads only to very little additional computational effort compared to a fixed terminal region. Namely, the only additional computation which has to be performed at each time step is determining $\alpha(t)$ via* (4.47), *which is a simple algebraic operation.* $\square$

Given the sets $\mathbb{X}^f(t)$, we can now define the sets $\overline{\mathbb{Y}}(t)$ such that Assumptions 4.4 and 4.5 are satisfied.

Proposition 4.34. *Suppose that for each $t \in \mathbb{I}_{\geq 0}$, the terminal region $\mathbb{X}^f(t)$ is given by* (4.46)–(4.47). *Furthermore, assume that $|h(x, \kappa^f(x))|_\mathbb{Y} \leq \gamma(E(x))$ for all $x \in \mathbb{X}^f(0)$ with $\gamma(s) := as^r$ and some constants $a, r > 0$. Then the choice*

$$\overline{\mathbb{Y}}(t) = bc^t B_1(0) \tag{4.49}$$

with $b := a\alpha_0^r$ and $c := (1 - \lambda_{\min}(Q)/\lambda_{\max}(P))^r$ satisfies Assumptions 4.4 and 4.5 with $\sigma(t) = b/(1-c)$ for all $t \in \mathbb{I}_{\geq 1}$.

Proof. Let $t' := \inf_{\tau \in \mathbb{I}_{\geq 0}, \alpha(\tau) = \alpha_{\max}} \tau$. The definition of $\mathbb{X}^f(t)$ in (4.46)–(4.47) yields that for all[10] $t \in \mathbb{I}_{[0,t'-1]}$ and all $x \in \mathbb{X}^f(t)$,

$$E(x) \leq \alpha(t) = \alpha_0(1 - \lambda_{\min}(Q)/\lambda_{\max}(P))^t.$$

---

[10]If $t' = +\infty$, then the following holds for all $t \in \mathbb{I}_{\geq 0}$.

Then, as $\mathbb{X}^f(t) \subseteq \mathbb{X}^f(0)$ for all $t \in \mathbb{I}_{[0,t'-1]}$, by assumption there exist constants $a, r > 0$ such that

$$|h(x, \kappa^f(x))|_\mathbb{Y} \leq \gamma(E(x)) \leq a\alpha_0^r (1 - \lambda_{\min}(Q)/\lambda_{\max}(P))^{rt} = bc^t$$

for all $x \in \mathbb{X}^f(t)$. Note that this inequality is also valid for all $t \in \mathbb{I}_{\geq t'+1}$, as for all such t we have $|h(x, \kappa^f(x))|_\mathbb{Y} = 0$ for all $x \in \mathbb{X}^f(t)$ due to the definition of t'. But this implies that Assumption 4.4 is satisfied if the sets $\overline{\mathbb{Y}}(t)$ are defined as in (4.49) with $b := a\alpha_0^r$ and $c := (1 - \lambda_{\min}(Q)/\lambda_{\max}(P))^r$. Furthermore, note that $c < 1$ (which follows from positive definiteness of P and Q) and $c \geq 0$ (which follows from (4.48) together with the fact that $E(x) \geq 0$ for all x). Hence for each $t \in \mathbb{I}_{\geq 0}$ we obtain

$$\sum_{k=0}^{t-1} \overline{\mathbb{Y}}(k) = B_1(0) \sum_{k=0}^{t-1} bc^t \subseteq B_1(0) \sum_{k=0}^{\infty} bc^t = \frac{b}{1-c} B_1(0). \tag{4.50}$$

But this implies that Assumption 4.5 is satisfied with $\sigma(t) = b/(1-c)$ for all $t \in \mathbb{I}_{\geq 1}$, which concludes the proof of the proposition. □

Remark 4.35. *In Proposition 4.34, assuming the existence of constants $a, r > 0$ such that $|h(x, \kappa^f(x))|_\mathbb{Y} \leq \gamma(E(x))$ for all $x \in \mathbb{X}^f(0)$ with $\gamma(s) := as^r$ is not a major restriction and is satisfied for many output functions h. Namely, as $h(x^*, \kappa^f(x^*)) \in \mathbb{Y}$, this assumption is satisfied if [11] $|h(x, \kappa^f(x))|_\mathbb{Y}$ is Lipschitz continuous (or, more general, Hölder continuous [12]) in x on the (compact) set $\mathbb{X}^f(0)$.*

Remark 4.36. *In case that $\alpha_{\max} > 0$, one could also use the (constant) terminal region $\mathbb{X}_f(t) = \{x \in \mathbb{R}^n : E(x) \leq \alpha_{\max}\}$ for all $t \in \mathbb{I}_{\geq 0}$, in which case Assumptions 4.4 and 4.5 are satisfied with $\overline{\mathbb{Y}}(t) = \{0\}$ and $\sigma(t) = 0$ for all $t \in \mathbb{I}_{\geq 0}$, as discussed above. However, using the time-varying terminal regions (4.46)–(4.47) results in a possibly larger feasible region $\mathbb{X}_N(0)$ in case that $\alpha_0 > \alpha_{\max}$. Furthermore, in order to retain the benefits of a larger terminal region, the tightening of the terminal region via (4.46)–(4.47) can be relaxed as follows. Namely, (4.47) can be applied only after some (arbitrary) finite time $\bar{t} \in \mathbb{I}_{\geq 1}$, and the constant terminal region $\mathbb{X}^f(t) = \mathbb{X}^f(0)$ is used for $t \in \mathbb{I}_{[0,\bar{t}-1]}$. Then, using $\overline{\mathbb{Y}}(t) = bc^{\max\{0, t-\bar{t}\}} B_1(0)$ instead of (4.49), the above analysis holds true in a similar way, and again Assumptions 4.4 and 4.5 (with some constant σ) are satisfied.*

Extension to time-varying output functions

We now show that the results of Theorem 4.29 can be extended to the case of time varying output functions $h(x, u, t)$. This will be important later in Chapter 5 on distributed economic MPC. To this end, we modify Assumption 2.7 as follows.

Assumption 4.7. *The function $h : \mathbb{X} \times \mathbb{U} \times \mathbb{I}_{\geq 0} \to \mathbb{R}^p$ is continuous and bounded and the set $\mathbb{Y} \subseteq \mathbb{R}^p$ is closed and convex.*

---

[11] If, for example, the set $\mathbb{Y}$ is given by $\mathbb{Y} = \mathbb{R}^p_{\leq 0}$ as in the following section, this is satisfied if $h(x, \kappa^f(x))$ is locally Lipschitz, respectively, Hölder continuous in x.

[12] A function $f : \mathbb{R}^n \to \mathbb{R}^m$ is Hölder continuous on a set $X \subseteq \mathbb{R}^n$ if $|f(x) - f(y)| \leq a|x - y|^b$ for all $x, y \in X$ and some $a, b > 0$.

Now let $\boldsymbol{\rho} : \mathbb{I}_{\geq 0} \to \mathbb{R}_{\geq 0}$ be any sequence such that $h(x, u, t+1) - h(x, u, t) \in \rho(t)B_1(0)$ for all $t \in \mathbb{I}_{\geq 0}$ and all $(x, u) \in \mathbb{Z}$. We then consider the following modifications of Assumptions 4.4 and 4.5, respectively.

Assumption 4.8. *For each* $t \in \mathbb{I}_{\geq 0}$, *the set* $\overline{\mathbb{Y}}(t)$ *is such that* $\{h(x, \kappa^f(x), t)\} \oplus N\rho(t)B_1(0) \subseteq$ $\mathbb{Y} \oplus \overline{\mathbb{Y}}(t)$ *for all* $x \in \mathbb{X}^f(t)$.

Assumption 4.9. *The output function* h *is given as* $h(x, u, t) := \hat{h}(x, u) + \varphi(x, u, t)$ *with* $|\varphi(x, u, t)| \leq \hat{\varphi}(t)$ *for all* $(x, u) \in \mathbb{Z}$ *and all* $t \in \mathbb{I}_{\geq 0}$, *and some* $\hat{\boldsymbol{\varphi}} : \mathbb{I}_{\geq 0} \to \mathbb{R}_{\geq 0}$. *Furthermore, there exists a sequence* $\hat{\boldsymbol{\sigma}} : \mathbb{I}_{\geq 0} \to \mathbb{R}_{\geq 0}$ *with* $\lim_{t \to \infty} \hat{\sigma}(t)/t = 0$ *such that* $\sum_{k=0}^{t-1}\left(\overline{\mathbb{Y}}(k) \oplus \hat{\varphi}(k)B_1(0)\right) \subseteq \hat{\sigma}(t)B_1(0)$ *for all* $t \in \mathbb{I}_{\geq 0}$.

We can now obtain the following result, which is a modification of Theorem 4.29.

Theorem 4.37. *Suppose that Assumptions 2.1, 2.3, 2.6, 4.3i) and ii), 4.7 and 4.8 are satisfied. Then the following holds for the closed-loop system* (2.6) *resulting from application of Algorithm 4.28:*

i) *If* $x_0 \in \mathbb{X}_N(0)$, *then Problem 4.23 is feasible for all* $t \in \mathbb{I}_{\geq 0}$, *and* (4.40) *is satisfied.*

ii) *If Assumption 4.5 is satisfied, then* (4.41) *holds.*

iii) *If Assumption 4.9 is satisfied, then* $\mathrm{Av}[\hat{h}(\boldsymbol{x}, \boldsymbol{u})] \subseteq \mathbb{Y}$.

Proof. Statement *i*) can be proven as in Theorem 4.29, with the following modification concerning recursive feasibility of constraint (4.36e). Consider again the candidate input and state sequences $\tilde{\boldsymbol{u}}(t+1)$ and $\tilde{\boldsymbol{x}}(t+1)$, respectively, from the proof of Theorem 4.29. Thanks to Assumption 4.8, we obtain

$$\sum_{k=0}^{N-1} h(\tilde{x}(k|t+1), \tilde{u}(k|t+1), t+1) \in \left\{\sum_{k=0}^{N-1} h(\tilde{x}(k|t+1), \tilde{u}(k|t+1), t)\right\} \oplus N\rho(t)B_1(0)$$

$$= \left\{\sum_{k=0}^{N-1} h(x^0(k|t), u^0(k|t), t) - h(x(t), u(t), t) + h(x^0(N|t), \kappa^f(x^0(N|t)), t)\right\} \oplus N\rho(t)B_1(0)$$

$$\subseteq \mathbb{Y}_t \oplus \left\{-h(x(t), u(t), t))\right\} \oplus \mathbb{Y} \oplus \overline{\mathbb{Y}}(t) = \mathbb{Y}_{t+1},$$

and hence Problem 4.23 is feasible for all $t \in \mathbb{I}_{\geq 0}$. Statement *ii*) of the theorem follows as in Theorem 4.29, exploiting the fact that h is bounded on $\mathbb{X} \times \mathbb{U} \times \mathbb{I}_{\geq 0}$ according to Assumption 4.7. Finally, for statement *iii*), we use $h(x, u, k) := \hat{h}(x, u) + \varphi(x, u, k)$ in the first sum of the left hand side of (4.43) to obtain

$$\sum_{k=0}^{t-1} \hat{h}(x(k), u(k)) + \sum_{k=0}^{N-1} h(x(k|t), u(k|t), t)$$

$$\in \mathbb{Y}_{00} \oplus (t+N)\mathbb{Y} \oplus \sum_{k=0}^{t-1} \overline{\mathbb{Y}}(k) \oplus B_1(0) \sum_{k=0}^{t-1} \hat{\varphi}(k). \tag{4.51}$$

Following the same calculations as in (4.44)–(4.45), we can conclude that

$$\lim_{n \to +\infty} \frac{\sum_{k=0}^{t_n-1} \hat{h}(x(k), u(k))}{t_n} \subseteq \lim_{n \to +\infty} \frac{\mathbb{Y}_{00} \oplus (t_n + N)\mathbb{Y} \oplus \hat{\sigma}(t_n)B_1(0)}{t_n} = \mathbb{Y}, \tag{4.52}$$

where the last equality again follows from the fact that $\mathbb{Y}_{00}$ is compact and $\lim_{t \to \infty} \hat{\sigma}(t)/t = 0$ according to Assumption 4.9. $\qquad\square$

Remark 4.38. *A close inspection of the respective proofs reveals that the results of Theorems 4.29 and 4.37 are still valid under the following modifications. Firstly, Problem 4.23 can be modified by replacing the term $\mathbb{Y} \oplus \overline{\mathbb{Y}}(t)$ in (4.38) with $\{h(x^0(N|t), \kappa^f(x^0(N|t)))\}$. Secondly, one can also replace the set $\mathbb{Y}_t$ in (4.38) with any set $\widetilde{\mathbb{Y}}_t \subseteq \mathbb{Y}_t$ such that $\sum_{k=0}^{N-1} h(x^0(k|t), u^0(k|t)) \in \widetilde{\mathbb{Y}}_t$ (respectively, $\sum_{k=0}^{N-1} h(x^0(k|t), u^0(k|t), t) \in \widetilde{\mathbb{Y}}_t$). These modifications can be used to reduce the size of the sets $\mathbb{Y}_t$, which can be desirable if the average constraints are used to enforce convergence of the closed-loop system, as smaller sets $\mathbb{Y}_t$ result in a faster convergence rate (compare Section 4.2.2 and Chapter 5). Thirdly, one can relax Assumption 4.4 by replacing $h(x, \kappa^f(x))$ for all $x \in \mathbb{X}^f(t)$ with $h(x^0(N|t), \kappa^f(x^0(N|t)))$; analogously, Assumption 4.8 can be relaxed by replacing $h(x, \kappa^f(x), t)$ for all $x \in \mathbb{X}^f(t)$ with $h(x^0(N|t), \kappa^f(x^0(N|t)), t)$.*

Remark 4.39. *Note that in contrast to Theorem 4.29, in Theorem 4.37 we did not establish the average performance bound (4.42). The reason for this is that if h is time-varying, also the set of optimal state/input equilibrium pairs S_{av}^* as defined in (2.16) (and hence (x^*, u^*)) is possibly time-varying. If h is such that (x^*, u^*) is constant for all t and in addition Assumption 4.3iii) is satisfied, then again (4.42) can be established as in Theorem 4.29. Furthermore, in Chapter 5 (compare Theorem 5.8), we show that (4.42) can also be established in case that h is such that the optimal state/input equilibrium pair asymptotically converges to some constant (x^*, u^*).*

4.2.2 Convergence in economic MPC with average constraints

As discussed above, the closed-loop system resulting from application of an economic MPC scheme is in general not necessarily convergent. However, for economic MPC without average constraints, it was shown in Theorem 2.10 that (strict) dissipativity cannot only be used to classify optimal steady-state operation of a system, but is also a sufficient condition for ensuring asymptotic stability of the closed-loop system. For economic MPC with average constraints, a first preliminary result concerning closed-loop convergence was obtained in (Angeli et al., 2012, Remark 6.5) in a setting with a terminal equality constraint. There, it was noted that if the system is strictly dissipative with respect to the supply rate $s(x, u) = \ell(x, u) - \ell(x^*, u^*) + \bar{\lambda}^T h(x, u)$ for some (arbitrary) $\bar{\lambda} \in \mathbb{R}^p_{\geq 0}$, then the closed-loop system satisfies

$$\liminf_{t \to \infty} |x(t) - x^*| = 0. \tag{4.53}$$

This result was proven by showing that the dissipativity condition implies that the system is suboptimally operated off steady-state (see Section 2.2.2), and combining this fact with the average performance result of Theorem 4.29, i.e., inequality (4.42). However, (4.53) only implies convergence of the closed-loop in a weak sense, and furthermore a Lyapunov-like stability analysis has not been available so far for economic MPC with average constraints. The main result of this section is to close this gap and prove asymptotic convergence of the closed-loop economic MPC solution to the optimal steady-state x^* by means of a Lyapunov-like analysis. To this end, we make the following assumption on the average constraint set $\mathbb{Y}$.

Assumption 4.10. *The average constraint set $\mathbb{Y}$ is given by $\mathbb{Y} = \mathbb{R}^p_{\leq 0}$.*

Note that this assumption is not a major restriction, as the output map h can be some general nonlinear function.[13] We can now state the following theorem, which establishes asymptotic convergence of the closed-loop system in economic MPC with asymptotic average constraints.

Theorem 4.40. *Let $x_0 \in \mathbb{X}_N(0)$, and suppose that Assumptions 2.1, 2.3, 2.6, 2.7, 4.3, 4.4 and 4.10 are satisfied, as well as Assumption 4.5 with a bounded sequence $\boldsymbol{\sigma}$. Furthermore, assume that system (2.1) is strictly dissipative on $\mathbb{Z}_{av}^0$ with respect to the supply rate $s(x, u) = \ell(x, u) - \ell(x^*, u^*) + \bar{\lambda}^T h(x, u)$ for some (arbitrary) $\bar{\lambda} \in \mathbb{R}_{\geq 0}^p$. Then the solution of the closed-loop system (2.6) resulting from application of Algorithm 4.28 asymptotically converges to x^*, i.e., $\lim_{t \to \infty} x(t) = x^*$.*

Proof. By Theorem 4.29, Problem 4.23 is recursively feasible, and hence the solution of the closed-loop system (2.6) resulting from application of Algorithm 4.28 is defined and exists for all $t \in \mathbb{I}_{\geq 0}$. Now denote by $L(x, u) := \ell(x, u) - \ell(x^*, u^*) + \lambda(x) - \lambda(f(x, u))$ the rotated stage cost and by $\widetilde{V}^f(x) := V^f(x) - V^f(x^*) + \lambda(x) - \lambda(x^*)$ the rotated terminal cost, where λ is the storage function from (2.12). Let $\widetilde{J}_N^0(x(t))$ denote the optimal value function of Problem 4.23 with ℓ and V^f in (4.37) replaced by L and $\widetilde{V}^f$, respectively. As was shown in (Amrit et al., 2011, Lemma 14), the minimizer of this modified optimization problem is identical to the minimizer of the original Problem 4.23, as the cost functions only differ by a constant term and the constraints are the same. In order to prove Theorem 4.40, we propose to use the following Lyapunov-like function, which we define along the solution of the resulting closed-loop system (2.6):

$$V(t) = \widetilde{J}_N^0(x(t)) + w(t) \qquad (4.54)$$

with

$$w(t) := \sup_{\tau \in \mathbb{I}_{\geq 0}} \sum_{k=t}^{t+\tau} \bar{\lambda}^T h(x(k), u(k)), \qquad (4.55)$$

where $h(x(\cdot), u(\cdot))$ is the output along the solution of the closed-loop system (2.6) from time t on.

The function w has the following properties. Firstly, we have that

$$w(t) \geq \bar{\lambda}^T h(x(t), u(t)) \geq \min_{(z,v) \in \mathbb{Z}} \bar{\lambda}^T h(z, v) =: \underline{w} > -\infty,$$

for all $t \in \mathbb{I}_{\geq 0}$, where the first inequality follows as $\tau = 0$ is allowed in the definition of w in (4.55), and the last inequality follows from compactness of $\mathbb{Z}$ and continuity of h. Secondly, from (4.36e), (4.39) and the fact that Assumption 4.5 is satisfied with a bounded sequence $\boldsymbol{\sigma}$, i.e., $\sigma(t) \leq \sigma_{\max} < \infty$ for all $t \in \mathbb{I}_{\geq 0}$, we obtain that the closed-loop system (2.6) satisfies

$$\sum_{k=0}^{N-1} h(x^0(k|t), u^0(k|t)) + \sum_{k=0}^{t-1} h(x(k), u(k)) \in \mathbb{Y}_{00} \oplus \sigma_{\max} B_1(0) \oplus (t + N)\mathbb{Y} \qquad (4.56)$$

---

[13]In particular, this includes the setup in Angeli et al. (2012), where $\mathbb{Y}$ was supposed to be a polyhedron given by $\mathbb{Y} = \{y \in \mathbb{R}^p : Ay \leq b\}$ with $A \in \mathbb{R}^{n_y \times p}$ and $b \in \mathbb{R}^{n_y}$ for some $n_y \in \mathbb{I}_{\geq 1}$. Namely, if necessary, one can define $\tilde{h}(x, u) := Ah(x, u) - b$ as a new auxiliary output function and then use the set $\mathbb{Y} = \mathbb{R}_{\leq 0}^{n_y}$ as in Assumption 4.10.

for all $t \in \mathbb{I}_{\geq 0}$. Hence, by Assumption 4.10, from (4.56) we obtain that for all $t \in \mathbb{I}_{\geq 0}$,

$$\sum_{k=0}^{t-1} h(x(k), u(k)) \leq y_{00} + \sigma' + y', \tag{4.57}$$

where y_{00}, σ' and y' are defined component-wise as

$$y_{00i} := \max_{y \in \mathbb{Y}_{00}} y_i, \quad \sigma'_i := \sigma_{\max}, \quad y'_i := N \max_{(x,u) \in \mathbb{Z}} -h_i(x, u)$$

for all $i \in \mathbb{I}_{[1,p]}$. Note that y_{00} and y' are finite due to compactness of $\mathbb{Y}_{00}$ and $\mathbb{Z}$ and continuity of h, respectively. Therefore, as $\bar{\lambda} \in \mathbb{R}^p_{\geq 0}$, from (4.57) we obtain that the closed-loop MPC solution satisfies, for all $t \in \mathbb{I}_{\geq 0}$,

$$\sum_{k=0}^{t-1} \bar{\lambda}^T h(x(k), u(k)) \leq \bar{\lambda}^T (y_{00} + \sigma' + y').$$

Thus it follows that $w(0) \leq \bar{\lambda}^T (y_{00} + \sigma' + y') < \infty$.

Next, consider the term $\widetilde{J}^0_N(x(t))$ in (4.54). As V^f is continuous, λ is bounded from below, and $\mathbb{X}^f(t)$ is contained in the compact set $\mathbb{Z}_\mathbb{X}$ for all $t \in \mathbb{I}_{\geq 0}$, there exists a constant $\underline{V}^f > -\infty$ such that $\underline{V}^f \leq \widetilde{V}^f(x)$ for all $x \in \mathbb{X}^f(t)$ and all $t \in \mathbb{I}_{\geq 0}$. Furthermore, we obtain that $L(x, u) \geq \rho(|x - x^*|) - \bar{\lambda}^T h(x, u) \geq \min_{(x,u) \in \mathbb{Z}} -\bar{\lambda}^T h(x, u) =: \underline{L} > -\infty$, where the first inequality follows from strict dissipativity and the last from continuity of h and compactness of $\mathbb{Z}$. Hence $N\underline{L} + \underline{V}^f \leq \widetilde{J}^0_N(x(t))$ for all $t \in \mathbb{I}_{\geq 0}$.

Combining the above, we obtain that for all $x_0 \in \mathbb{X}_N(0)$, the function V satisfies $V(0) \leq \bar{\lambda}^T (y_{00} + \sigma' + y') + \widetilde{J}^0_N(x_0) < \infty$ and $V(t) \geq \underline{w} + N\underline{L} + \underline{V}^f > -\infty$ for all $t \in \mathbb{I}_{\geq 0}$. Now consider the evolution of V along the solution of the closed-loop system (2.6) resulting from application of Algorithm 4.28. First, by considering the candidate input and state sequences $\tilde{\boldsymbol{u}}(t+1)$ and $\tilde{\boldsymbol{x}}(t+1)$, respectively, from the proof of Theorem 4.29, we obtain that

$$\begin{aligned}
\widetilde{J}^0_N(x(t+1)) - \widetilde{J}^0_N(x(t)) \leq{} & L(x^0(N|t), \kappa^f(x^0(N|t))) - L(x(t), u(t)) \\
& + \widetilde{V}^f(f(x^0(N|t), \kappa^f(x^0(N|t)))) - \widetilde{V}^f(x^0(N|t)) \\
\leq{} & -L(x(t), u(t)), \tag{4.58}
\end{aligned}$$

where the last inequality follows from Assumption 4.3(iii) and the definitions of L and $\widetilde{V}^f$ (see (Amrit et al., 2011, Lemma 9)). With this, we obtain

$$\begin{aligned}
V(t+1) - V(t) ={} & \widetilde{J}^0_N(x(t+1)) - \widetilde{J}^0_N(x(t)) + w(t+1) - w(t) \\
\overset{(4.58)}{\leq}{} & -L(x(t), u(t)) + w(t+1) - w(t) \\
\leq{} & -\rho(|x(t) - x^*|) + \bar{\lambda}^T h(x(t), u(t)) + w(t+1) - w(t) \\
\overset{(4.55)}{=}{} & -\rho(|x(t) - x^*|) + \bar{\lambda}^T h(x(t), u(t)) \\
& + \sup_{\tau \in \mathbb{I}_{\geq 0}} \sum_{k=t+1}^{t+1+\tau} \bar{\lambda}^T h(x(k), u(k)) - \sup_{\tau \in \mathbb{I}_{\geq 0}} \sum_{k=t}^{t+\tau} \bar{\lambda}^T h(x(k), u(k)) \\
={} & -\rho(|x(t) - x^*|) + \sup_{\tau \in \mathbb{I}_{\geq 1}} \sum_{k=t}^{t+\tau} \bar{\lambda}^T h(x(k), u(k)) - \sup_{\tau \in \mathbb{I}_{\geq 0}} \sum_{k=t}^{t+\tau} \bar{\lambda}^T h(x(k), u(k)) \\
\leq{} & -\rho(|x(t) - x^*|), \tag{4.59}
\end{aligned}$$

where the second inequality follows from strict dissipativity of the system with respect to the supply rate $s(x, u) = \ell(x, u) - \ell(x^*, u^*) + \bar{\lambda}^T h(x, u)$. Summarizing the above, the sequence $V(t)$ is non-increasing in t, bounded from below and $V(0)$ is finite; hence it converges. But this implies according to (4.59) that $\sum_{t=0}^{\infty} \rho(|x(t) - x^*|)$ converges, which by the fact that $\rho \in \mathcal{K}_{\infty}$ in turn implies that $\lim_{t \to \infty} x(t) = x^*$, i.e., the closed-loop system (2.6) resulting from application of Algorithm 4.28 asymptotically converges to x^* as claimed. $\qquad\square$

Remark 4.41. *In Theorem 4.40, relaxing Assumption 2.6 to Assumption 2.2 in order to also allow for an unbounded constraint set $\mathbb{Z}$ can be done in a similar way as discussed in Remarks 2.11 and 4.30, respectively. Namely, if $\mathbb{Z}$ is unbounded, the proof of Theorem 4.40 can be modified as follows. In order to establish the lower and upper bounds for $w(t)$ and $w(0)$, respectively, one would in addition need to assume that h is bounded from below on $\mathbb{Z}$. To obtain the lower bound for $\tilde{J}_N^f(x)$, one would need to require that V^f is bounded from below on $\cup_{t \in \mathbb{I}_{\geq 0}} \mathbb{X}^f(t) \subseteq \mathbb{Z}_\mathbb{X}$ and h is in addition also bounded from above on $\mathbb{Z}$; the latter could alternatively be replaced by requiring that ℓ is bounded from below on $\mathbb{Z}$ and λ is bounded on $\mathbb{X}_{av}^0$.*

Discussion

In the following, we comment on various notable properties of Theorem 4.40 and its proof, before considering an example illustrating its results.

First, note that the function V (defined in (4.54)), which is used in the proof of Theorem 4.40, is different from the Lyapunov function used in economic MPC without average constraints (Amrit et al., 2011; Angeli et al., 2012). Namely, in these references, $\tilde{J}_N^0(x)$ is used as a Lyapunov function, while here we need the additional term w defined by (4.55) in order to be able to conclude the decaying of V via (4.59).

Second, in Theorem 4.40, requiring Assumption 4.5 to hold with a bounded sequence σ is crucial for establishing boundedness of $w(0)$ and hence $V(0)$. Namely, if Assumption 4.5 is only satisfied for some unbounded σ, then no finite upper bound for $w(0)$ can be found via (4.56)–(4.57). As discussed above, the sets $\bar{\mathbb{Y}}(t)$ can always be chosen such that Assumption 4.5 is satisfied with bounded (constant) σ, if, for example, (i) a terminal equality constraint is used, or (ii) $h(x, \kappa^f(x)) \in \mathbb{Y}$ for all $x \in \mathbb{X}^f$, or (iii) if the tightening of the terminal region as proposed in Section 4.2.1 is applied, i.e., $\mathbb{X}^f(t)$ and $\bar{\mathbb{Y}}(t)$ are calculated according to (4.46)–(4.47) and (4.49), respectively.

Finally, while in Theorem 4.40 we showed asymptotic convergence of the closed-loop solution to x^*, note that x^* is not necessarily asymptotically stable (i.e., lacks the Lyapunov stability property). This is in contrast to the setting without average constraints, where strict dissipativity leads to an asymptotically stable equilibrium x^* (compare Theorem 2.10). Namely, the average constraints allow the system to initially "spend time" in a region of the state-space where it is not allowed on average. This means that even when starting at the optimal steady-state x^*, the closed-loop MPC trajectory might not stay there (see Example 4.42 below). On the other hand, we note that if the strict dissipativity condition of Theorem 4.40 is strengthened to hold with $\bar{\lambda} = 0$ (and x^* lies in the interior of $\mathbb{X}^f(0)$), then again asymptotic stability of x^* can be established as was done in Theorem 2.10 for systems without average constraints. In fact, $w(t) \equiv 0$ in this case, and hence Theorem 4.40 is valid without requiring Assumption 4.5 to hold.

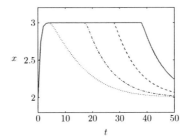

Figure 4.3: Closed-loop state sequences in Example 4.42 with $y_{00} = 20$ (dotted), $y_{00} = 40$ (dash-dotted), $y_{00} = 60$ (dashed), and $y_{00} = 80$ (solid).

Example 4.42. *Consider the system (4.20) of example 4.13 with state and input constraint set $\mathbb{Z} = [-10, 10]^2$ and average constraint of the form (2.14) with $h(x, u) = 2x + u - 5$ and $\mathbb{Y} = \mathbb{R}_{\leq 0}$. The stage cost is given by $\ell(x, u) = (x - 3)^2 + u^2$, and hence the set of optimal state/input equilibrium pairs as defined in (2.16) is given by $S_{av}^* = \{(2, 1)\}$ with $\ell(2, 1) = 2$. The terminal cost V^f and the sets $\mathbb{X}^f(t)$ and $\overline{\mathbb{Y}}(t)$ were calculated as described in Section 4.2.1 such that Assumptions 4.3–4.5 (the latter with bounded $\boldsymbol{\sigma}$) are satisfied. Furthermore, one can show that system (4.20) is strictly dissipative with respect to the supply rate $s(x, u) = \ell(x, u) - \ell(x^*, u^*) + \bar{\lambda}^T h(x, u)$ with $\bar{\lambda} = 1$ and with storage function $\lambda(x) = (3/2)x$. Figure 4.3 shows closed-loop state sequences resulting from application of Algorithm 4.28 with prediction horizon $N = 10$, where $\mathbb{Y}_{00}$ in (4.38) is given by $\mathbb{Y}_{00} = \{y \in \mathbb{R} : |y| \leq y_{00}\}$ with different values of y_{00}. One can see that the closed-loop system converges to x^*. However, x^* is not a Lyapunov stable equilibrium point. Namely, although $x_0 = x^* = 2$, the closed-loop first moves away from x^* and approaches $x = 3$, before finally converging to $x^* = 2$. In fact, it turns out that without the average constraint, the optimal state/input equilibrium pair would be given by $(x, u) = (3, 1)$ with $\ell(3, 1) = 1 < \ell(2, 1)$, and the system would be strictly dissipative with respect to the supply rate $s(x, u) = \ell(x, u) - \ell(3, 1)$ and with storage function $\lambda(x) = (2/3)x$. Hence the closed-loop system would asymptotically converge to $x = 3$ if no average constraint was present. Figure 4.3 can now be interpreted as follows. The average constraint initially allows the system to converge to $x = 3$, before forcing it to leave this steady-state again and to converge to $x^* = 2$, which is the best steady-state also fulfilling the average constraint. As discussed in Remark 4.25, the amount of time the system is allowed to stay in a vicinity of $x = 3$ can be tuned by choosing the parameter y_{00} in $\mathbb{Y}_{00}$ accordingly (see Figure 4.3).*

Using average constraints to enforce convergence in economic MPC

As already discussed above, in case that the dissipativity condition of Theorem 2.10 and 4.40 is not satisfied, the closed-loop system resulting from application of Algorithm 2.3 and 4.28, respectively, is in general not convergent. For some applications, such a behavior might be acceptable, while for others, closed-loop convergence is a critical requirement which cannot be abandoned for the benefit of enhancing performance (compare Angeli et al. (2011)). In order to enforce convergence of the closed-loop, different methods can be used. In Angeli et al. (2012), the authors propose to modify the stage cost ℓ by an additional positive definite term resulting in the system being dissipative. On the other hand, in (Angeli et al., 2011; Müller et al., 2014f) closed-loop convergence was enforced by means of imposing an additional (auxiliary) average constraint on the system, in which case the stage cost ℓ needs not be modified, i.e., the original (economic) cost function can be used. Namely, in (Müller et al., 2014f, Proposition 3) it was shown that when, e.g., the additional average constraint output $h(x,u) = |x - x^*| + |u - u^*|$ is used, there exists some $\bar{\lambda} \geq 0$ such that the dissipativity condition of Theorem 4.40 is satisfied, and hence closed-loop convergence can be established via Theorem 4.40. Alternatively, in (Angeli et al., 2011; Müller et al., 2014f), an additional zero-moment average constraint was used, i.e., $h(x,u) = (x - x^*)^q$ for some even $q \in \mathbb{I}_{\geq 2}$ (powers are intended componentwise), or $h(x,u) = |x - x^*|^q$ for some $q > 0$, together with $\mathbb{Y}$ given as in Assumption 4.10. Then Theorem 4.29 can be used to ensure that $Av[h(\boldsymbol{x}, \boldsymbol{u})] = \{0\}$ for the closed-loop system, from where essential convergence[14] of the closed-loop system to x^* follows (see Angeli et al. (2011); Müller et al. (2014f)). In case that Assumption 4.5 is satisfied with a bounded sequence $\boldsymbol{\sigma}$, asymptotic instead of only essential convergence can be established. For further details as well as examples, the interested reader is referred to (Angeli et al., 2011; Müller et al., 2014f).

4.2.3 Economic MPC with transient average constraints

In the previous sections, we have considered economic MPC with asymptotic average constraints of the form (2.14). In this section, we propose an economic MPC scheme which ensures satisfaction of *transient* average constraints, i.e., constraints imposed on input and state variables averaged over some finite time period T. Similar to what was discussed at the beginning of Section 2.2.2 for asymptotic average constraints, also transient average constraints are natural and of particular interest (and have to be taken into account online) in the context of economic MPC, as the closed-loop system is not necessarily convergent. However, the same is now also true for the initial phase (i.e., before converging) in standard tracking MPC.

To be precise, for some given finite time period $T \in \mathbb{I}_{\geq 1}$, the transient average constraints imposed on the system (2.1) are such that the following has to be satisfied for all $t \in \mathbb{I}_{\geq 0}$:

$$\sum_{k=t}^{t+T-1} \frac{h(x(k), u(k))}{T} \leq 0, \tag{4.60}$$

where $h(x,u) \in \mathbb{R}^p$ is again the average constraint output variable. This means that each finite-time (transient) average of h with period T has to be contained in the set $\mathbb{Y} = \mathbb{R}^p_{\leq 0}$, whereas this was only required to hold asymptotically in case of the asymptotic average constraints (2.14). The set of feasible, respectively, optimal state/input equilibrium pairs can now again be computed as in (2.15) and (2.16), respectively (with $\mathbb{Y} = \mathbb{R}^p_{\leq 0}$). As above, (x^*, u^*) denotes an arbitrary element of the set S^*_{av} as defined in (2.16). Note that $T = 1$ in (4.60) corresponds to standard pointwise-in-time state and input constraints (2.2) (compare also Remark 4.45 below). Hence in the following, we are interested in the case of $T \in \mathbb{I}_{\geq 2}$. To this end, given some $T \in \mathbb{I}_{\geq 2}$, we propose the following optimization problem to be solved at each time $t \in \mathbb{I}_{\geq 0}$ with measured state $x(t)$:

---

[14]See Appendix A.2 for the definition of essential convergence.

Problem 4.43.

$$\underset{\{u(0|t),\ldots,u(N-1|t)\}}{minimize} \; J_N(x(t), \boldsymbol{u}(t)) \tag{4.61}$$

subject to

$$x(k+1|t) = f(x(k|t), u(k|t)), \quad k \in \mathbb{I}_{[0,N+T-3]}, \tag{4.62a}$$

$$x(0|t) = x(t), \tag{4.62b}$$

$$(x(k|t), u(k|t)) \in \mathbb{Z}, \quad k \in \mathbb{I}_{[0,N-1]}, \tag{4.62c}$$

$$x(N|t) \in \mathbb{X}^f(t), \tag{4.62d}$$

$$\sum_{k=t-T+i}^{t-1} h(x(k), u(k)) + \sum_{k=0}^{i-1} h(x(k|t), u(k|t)) \le \varphi(t+i-T-N+1), \quad i \in \mathbb{I}_{[\max\{1,T-t\},T]}, \tag{4.62e}$$

$$\sum_{k=j}^{j+T-1} h(x(k|t), u(k|t)) \le \varphi(t+j-N+1), \quad j \in \mathbb{I}_{[1,N-1]}, \tag{4.62f}$$

$$u(k|t) = \kappa^f(x(k|t)), \quad k \in \mathbb{I}_{[N,N+T-2]}, \tag{4.62g}$$

with $J_N(x(t), \boldsymbol{u}(t))$ as defined in (4.37).

Compared to Problems 2.2 and 4.23, the main novel feature of Problem 4.43 are the constraints (4.62e)–(4.62f). Each of these constraints consists of a sum of T addends, and in total $(N+T-1)p$ such constraints are imposed via (4.62e)–(4.62f). In (4.62e), the terms in the first sum[15] are the output values along the past closed-loop solution up to time $t-1$, and $\boldsymbol{u}(t) := \{u(0|t),\ldots,u(N+T-2|t)\}$ and $\boldsymbol{x}(t) := \{x(0|t),\ldots,x(N+T-2|t)\}$ denote again predicted input and corresponding state sequences, respectively. Note that while predicted inputs and states up to $k = N + T - 2$ are now needed in (4.62f), the number of free input variables is *not* increased in Problem 4.43, i.e., the minimization is still done over the first N elements $\{u(0|t),\ldots,u(N-1|t)\}$, as was the case in Problems 2.2 and 4.23. For $k \ge N$, the predicted inputs and states are fixed by (4.62g) together with (4.62a), respectively. Furthermore, the (possibly time-varying) terminal region $\mathbb{X}^f$ and the sequence $\boldsymbol{\varphi} : \mathbb{I}_{\ge -N+1} \to \mathbb{R}^p$ will be specified later. As before, denote the minimizer of Problem 4.43 by $\boldsymbol{u}^0(t) := \{u^0(0|t),\ldots,u^0(N+T-2|t)\}$, the corresponding state sequence by $\boldsymbol{x}^0(t) := \{x^0(0|t),\ldots,x^0(N+T-2|t)\}$, and by $\mathbb{X}_N(t)$ the set of all states $x \in \mathbb{X}$ such that Problem 4.43 (with $x(t) = x$) has a solution.

Remark 4.44. *For ensuring satisfaction of asymptotic average constraints, p additional constraints (4.36e) were needed in Problem 4.23. On the other hand, in order to guarantee fulfillment of transient average constraints, $(N + T - 1)p$ additional constraints (4.62e)–(4.62f) are needed in Problem 4.43. Nevertheless, we again emphasize that the number of free input variables is the same for Problems 2.2, 4.23 and 4.43.*

Remark 4.45. *As already discussed above, $T = 1$ in (4.60) corresponds to standard pointwise-in-time state and input constraints (2.2). Indeed, in this case, Problem[16] 4.43*

---

[15]Recall that by convention, the empty sum is zero (for $i = T$).

[16]For $T = 1$, in (4.62a) one would need $k \in \mathbb{I}_{[0,N-1]}$ instead of $k \in \mathbb{I}_{[0,N+T-3]}$, as predicted states up to $k = N$ are needed (for the terminal constraint 4.62d).

would correspond to the standard Problem 2.2. Namely, for $T = 1$, (4.62g) disappears and (4.62e)–(4.62f) constitute some (additional) pointwise-in-time constraints (i.e., each constraint of (4.62e)–(4.62f) consists of only one term), which can be absorbed into (4.62c).

Remark 4.46. *We note that an alternative way to deal with transient average constraints would be to define the extended state*

$$z(t) := [x(t)^T \ \dots \ x(t-T+1)^T \ u(t-1)^T \ \dots \ u(t-T+1)^T]^T$$

and then use a (standard) economic MPC scheme for the corresponding extended system including standard pointwise-in-time constraints only. However, this alternative formulation has the following drawbacks. First, the terminal constraint for the extended system would in fact require that $x(N-T+1|t), \dots, x(N|t)$ are close to the optimal steady-state x^*, while via (4.62d) this is only required for $x(N|t)$. Furthermore, the terminal cost V_f would then be a function of $x(N-T+1|t), \dots, x(N|t)$ and $u(N-T+1|t), \dots, u(N-1|t)$ instead of $x(N|t)$ only, which means that a different cost than the original (economic) cost ℓ is used over this time horizon. Finally, applying stability analysis results for standard economic MPC to the extended system would in general result in a stricter dissipativity requirement than the one used in Theorem 4.53 below (involving the multiplier $\bar{\lambda}$). For these reasons, we propose to define the economic MPC scheme for the original system and take care of the transient average constraints by means of (4.62e)–(4.62f).

Economic MPC with exact satisfaction of transient average constraints

We first examine under what conditions an economic MPC scheme based on Problem 4.43 results in (exact) satisfaction of the transient average constraints (4.60), before considering the case where this can only be achieved approximately. For the first case, we assume that there exists a constant terminal region $\mathbb{X}^f$ satisfying Assumption 4.3 as well as the following.

Assumption 4.11. *The terminal region $\mathbb{X}^f$ and the local auxiliary control law κ^f are such that $h(x, \kappa^f(x)) \leq 0$ for all $x \in \mathbb{X}^f$.*

We can now define the economic model predictive control scheme including transient average constraints as follows.

Algorithm 4.47 (Economic MPC with transient average constraints). *Consider system (2.1). At each time $t \in \mathbb{I}_{\geq 0}$, measure the state $x(t)$, solve Problem 4.43 with constant terminal region $\mathbb{X}^f$ and $\varphi(t) = 0$ for all $t \in \mathbb{I}_{\geq -N+1}$, and apply the control input $u(t) := u^0(0|t)$.*

Remark 4.48. *In Algorithm 4.47, constraint (4.62f) in Problem 4.43 could be replaced by*

$$\sum_{k=j}^{\min\{j+T-1,N-1\}} h(x(k|t), u(k|t)) \leq 0, \quad j \in \mathbb{I}_{[1,N-1]}, \tag{4.63}$$

without changing the properties of the algorithm which are stated below. Then, one would not have to consider predicted input and state variables up to $k = N + T - 2$, but only as usual up to $k = N - 1$ and $k = N$, respectively, and hence (4.62g) could be omitted.

However, this alternative formulation is more conservative. Namely, as $x(k|t) \in \mathbb{X}^f$ and $u(k|t) = \kappa^f(x(k|t))$ for all $k \in \mathbb{I}_{[N,N+T-2]}$ (which is ensured by (4.62d), (4.62a) and (4.62g) together with Assumption 4.3ii)), by Assumption 4.11 it follows that $h(x(k|t), u(k|t)) \leq 0$ for all $k \in \mathbb{I}_{[N,N+T-2]}$, and hence requiring (4.62f) to hold is less restrictive than (4.63).

We can now state the following result, which shows that the closed-loop system resulting from application of Algorithm 4.47 satisfies the transient average constraints (4.60).

Theorem 4.49. *Let $x_0 \in \mathbb{X}_N(0)$, and suppose that Assumptions 2.1, 2.3, 2.6, 2.7, 4.3 and 4.11 are satisfied. Then Problem 4.43 with constant terminal region $\mathbb{X}^f$ and $\varphi \equiv 0$ is feasible for all $t \in \mathbb{I}_{\geq 0}$, and the closed-loop system (2.6) resulting from application of Algorithm 4.47 satisfies the pointwise-in-time constraints (4.40), the transient average constraints (4.60), and the asymptotic average performance bound (4.42).*

Proof. The proof of recursive feasibility is as usual by induction. Suppose that Problem 4.43 is feasible at time t. Then, consider at time $t + 1$ the candidate input sequence $\tilde{\boldsymbol{u}}(t+1) := \{u^0(1|t), \dots, u^0(N+T-2|t), \kappa^f(\tilde{x}(N+T-2|t+1))\}$ with corresponding state sequence $\tilde{\boldsymbol{x}}(t+1) := \{x^0(1|t), \dots, x^0(N+T-2|t), f(x^0(N+T-2|t), \kappa^f(x^0(N+T-2|t)))\}$. For these candidate input and state sequences, constraints (4.62a)–(4.62d) and (4.62g) are satisfied, which follows as usual from the fact that they were satisfied at time t and Assumption 4.3. Now consider constraint (4.62e) for some $i \in \mathbb{I}_{[\max\{1,T-t\},T-1]}$. Evaluated for the given candidate input and state sequences, the left-hand side reads

$$\sum_{k=t+1-T+i}^{t} h(x(k), u(k)) + \sum_{k=0}^{i-1} h(\tilde{x}(k|t+1), \tilde{u}(k|t+1))$$

$$= \sum_{k=t+1-T+i}^{t} h(x(k), u(k)) + \sum_{k=0}^{i-1} h(x^0(k+1|t), u^0(k+1|t))$$

$$= \sum_{k=t-T+(i+1)}^{t-1} h(x(k), u(k)) + \sum_{k=0}^{(i+1)-1} h(x^0(k|t), u^0(k|t)).$$

But this means that the i-th constraint of (4.62e) at time $t+1$ corresponds to the $(i+1)$-st constraint of (4.62e) at time t, which was satisfied by assumption. Hence constraint (4.62e) is satisfied at time $t + 1$ for all $i \in \mathbb{I}_{[\max\{1,T-t\},T-1]}$. Similarly, one can show that constraint (4.62e) with $i = T$ at time $t + 1$ evaluated for the candidate input and state sequences corresponds to constraint (4.62f) with $j = 1$ at time t, and that for each $j \in \mathbb{I}_{[1,N-2]}$, constraint (4.62f) at time $t + 1$ evaluated for the candidate input and state sequences corresponds to the $(j + 1)$-st constraint of (4.62f) at time t. Finally, when evaluated for the given candidate input and state sequences, constraint (4.62f) with $j = N-1$ only contains terms $h(x, \kappa^f(x))$ with $x \in \mathbb{X}^f$, which is the case due to (4.62g) and (4.62a) and the fact that $\tilde{x}(N-1|t+1) = x^0(N|t) \in \mathbb{X}^f$. But then, by Assumption 4.11 and the fact that $\varphi \equiv 0$, constraint (4.62f) with $j = N-1$ is also satisfied at time $t+1$. Hence the candidate input and state sequences $\tilde{\boldsymbol{u}}(t+1)$ and $\tilde{\boldsymbol{x}}(t+1)$, respectively, satisfy the constraints (4.62), which by induction implies that Problem 4.43 is feasible for all $t \in \mathbb{I}_{\geq 0}$.

Satisfaction of the pointwise-in-time constraints (4.40) for the closed-loop system (2.6) again follows by definition of the input $u(t)$ in Algorithm 4.47. Fulfillment of the transient average constraints (4.60) for all $t \in \mathbb{I}_{\geq 0}$ follows directly from the first constraint of (4.62e) (i.e., with $i = 1$) and the definition of the receding horizon control law. Namely, from (4.62e) with $i = 1$ (this constraint applies for all $t \geq T-1$), the definition of $u(t)$ in Algorithm 4.47 and the fact that $\varphi \equiv 0$, we obtain that along the closed-loop system (2.6) the following is satisfied for all $t \geq T - 1$:

$$\sum_{k=t-T+1}^{t-1} h(x(k), u(k)) + h(x^0(0|t), u^0(0|t)) = \sum_{k=t-T+1}^{t} h(x(k), u(k)) \leq 0.$$

By an index shift and considering that $T > 0$, this inequality is equivalent to the fact that (4.60) is satisfied for all $t \in \mathbb{I}_{\geq 0}$. Finally, the asymptotic average performance bound (4.42) follows as in Theorem 4.29. □

Remark 4.50. *Satisfaction of the transient average constraints (4.60) with some $T \in \mathbb{I}_{\geq 1}$ immediately implies that also transient average constraints (4.60) with period kT are satisfied, for each $k \in \mathbb{I}_{\geq 1}$. Furthermore, also the asymptotic average constraint (2.14) with $\mathbb{Y} = \mathbb{R}^p_{\leq 0}$ is satisfied in case that $\mathbb{Z}$ is compact, as for each $t \in \mathbb{I}_{\geq 0}$*

$$\sum_{k=0}^{t} \frac{h(x(k), u(k))}{t+1} = \sum_{i=0}^{\lfloor (t+1)/T \rfloor - 1} \sum_{k=iT}^{(i+1)T-1} \frac{h(x(k), u(k))}{t+1} + \sum_{k=\lfloor (t+1)/T \rfloor T}^{t} \frac{h(x(k), u(k))}{t+1}$$

$$\overset{(4.60)}{\leq} \sum_{k=\lfloor (t+1)/T \rfloor T}^{t} \frac{h(x(k), u(k))}{t+1} \qquad (4.64)$$

Hence, as the sum in (4.64) contains at most $T-1$ terms and furthermore h is continuous and $\mathbb{Z}$ compact, we obtain that

$$\limsup_{t \to \infty} \sum_{k=0}^{t} \frac{h(x(k), u(k))}{t+1} \leq \limsup_{t \to \infty} \sum_{k=\lfloor (t+1)/T \rfloor T}^{t} \frac{h(x(k), u(k))}{t+1} = 0, \qquad (4.65)$$

i.e., the asymptotic average constraint (2.14) with $\mathbb{Y} = \mathbb{R}^p_{\leq 0}$ is satisfied.

Considering Remark 4.50, we arrive at the following corollary.

Corollary 4.51. *Suppose that the conditions of Theorem 4.49 hold. Then for the closed-loop system (2.6) resulting from application of Algorithm 4.47, transient average constraints of the form (4.60) with period kT are satisfied for each $k \in \mathbb{I}_{\geq 1}$. Furthermore, the asymptotic average constraint (2.14) with $\mathbb{Y} = \mathbb{R}^p_{\leq 0}$ is satisfied.*

Remark 4.52. *In Theorem 4.49, relaxing Assumption 2.6 to Assumption 2.2, i.e., allowing for an unbounded constraint set $\mathbb{Z}$, can be done as discussed in the paragraph below Theorem 2.5; note that these modifications are only necessary for the average performance bound (4.42). For the considerations of Remark 4.50, one would then in addition need to ensure that the second term of (4.65) is less or equal than zero. This is, e.g, the case if h is bounded from above or from below on $\mathbb{Z}$. While the former is immediate to see, the latter is true due to the fact that $\limsup_{t \to \infty} \sum_{k=\lfloor (t+1)/T \rfloor T}^{t} h(x(k), u(k))$ can only be unbounded from above if h is unbounded from below, as $\sum_{k=\lfloor (t+1)/T \rfloor T}^{\lfloor (t+1)/T \rfloor (T+1)-1} h(x(k), u(k)) \leq 0$ for all $t \in \mathbb{I}_{\geq 0}$ by satisfaction of the transient average constraints. Hence boundedness of h from below on $\mathbb{Z}$ also implies that the second term of (4.65) is less or equal than zero.*

In Section 4.2.2, we have shown that in case of asymptotic average constraints and under a certain dissipativity condition, the closed-loop system asymptotically converges to the optimal steady-state x^*. The following result shows that the same holds true in the setting of transient average constraints. To this end, let the set $\mathbb{Z}_{av}^0$ be defined as in (2.17), but with the asymptotic average constraint $Av[h(\boldsymbol{z}, \boldsymbol{v})] \subseteq \mathbb{Y}$ replaced by the transient one $(1/T) \sum_{k=t}^{t+T-1} h(z(k), v(k)) \leq 0$ for all $t \in \mathbb{I}_{\geq 0}$.

Theorem 4.53. *Suppose that the conditions of Theorem 4.49 hold, and that system (2.1) is strictly dissipative on $\mathbb{Z}_{av}^0$ with respect to the supply rate $s(x,u) = \ell(x,u) - \ell(x^*,u^*) + \bar{\lambda}^T h(x,u)$ for some (arbitrary) $\bar{\lambda} \in \mathbb{R}_{\geq 0}^p$. Then the solution of the closed-loop system (2.6) resulting from application of Algorithm 4.47 asymptotically converges to x^*, i.e., $\lim_{t\to\infty} x(t) = x^*$.*

Proof. The proof of Theorem 4.53 follows the lines of the proof of Theorem 4.40. The only difference is calculating the upper bound for $w(0)$, with w as defined in (4.55). Using similar calculations as in (4.64) together with the fact that $\bar{\lambda} \in \mathbb{R}_{\geq 0}^p$, we obtain that for each $t \in \mathbb{I}_{\geq 0}$ and each $\tau \in \mathbb{I}_{\geq 0}$

$$\sum_{k=t}^{t+\tau} \bar{\lambda}^T h(x(k), u(k)) \leq \sum_{k=t+\lfloor (\tau+1)/T \rfloor T}^{t+\tau} \bar{\lambda}^T h(x(k), u(k)).$$

Again, the sum on the right hand side consists of at most $T-1$ terms, and hence for each $t \in \mathbb{I}_{\geq 0}$ we have

$$w(t) \leq (T-1) \max_{(x,u)\in\mathbb{Z}} \bar{\lambda}^T h(x,u) < \infty, \tag{4.66}$$

where the last inequality follows from continuity of h and compactness of $\mathbb{Z}$. $\qquad\square$

Remark 4.54. *In Theorem 4.53, relaxing Assumption 2.6 to Assumption 2.2, i.e., allowing for an unbounded constraint set $\mathbb{Z}$, can be done as discussed in Remark 4.41.*

Economic MPC with approximate satisfaction of transient average constraints

In case that Assumption 4.11 is not satisfied, the above results cannot readily be applied. Namely, the shifted candidate input and state sequences at time $t+1$ used in the proof of Theorem 4.49 do not necessarily satisfy (4.62f) with $\varphi \equiv 0$, as now there exist $x \in \mathbb{X}^f$ with $h(x, \kappa^f(x)) > 0$. Nevertheless, we will show in the following that the proposed economic MPC scheme can be modified such that a slightly relaxed form of the transient average constraints (4.60) can be ensured, namely that for all $t \in \mathbb{I}_{\geq 0}$

$$\sum_{k=t}^{t+T-1} \frac{h(x(k), u(k))}{T} \leq \psi(t), \tag{4.67}$$

where the sequence $\boldsymbol{\psi} : \mathbb{I}_{\geq 0} \to \mathbb{R}_{\geq 0}^p$ is (componentwise) nonincreasing and decaying to zero as $t \to \infty$. This means that the violation of the original transient average constraints (4.60) is upper-bounded by the term $\psi(t)$, which decays to zero as $t \to \infty$. To this end, in the following we consider again the time-varying terminal regions as defined by (4.46)–(4.47) based on Assumption 4.6. Recall that as discussed in Section 4.2.1, $\alpha_0 \leq \alpha_{\max}$ implies that $\mathbb{X}_f(t) = \mathbb{X}_f(0)$ for all $t \in \mathbb{I}_{\geq 0}$ and Assumption 4.11 is satisfied, i.e., Theorem 4.49 can be applied. Otherwise, the definition of $\mathbb{X}_f(t)$ in (4.46)–(4.47) means that the time-varying terminal region is gradually tightened such that ultimately (at least asymptotically) Assumption 4.11 is satisfied. Furthermore, we note that as h is continuous, it follows that if also κ^f is continuous with $h(x^*, \kappa^f(x^*)) \leq 0$, then there exists a function $\gamma : \mathbb{R}_{\geq 0} \to \mathbb{R}^p$ such that

$$h(x, \kappa^f(x)) \leq \gamma(E(x)) \tag{4.68}$$

for all $x \in \mathbb{X}_f(0)$, where $\gamma(r) = [\gamma_1(r) \ \ldots \ \gamma_p(r)]^T$ with $\gamma_i \in \mathcal{K}$ for all $i \in \mathbb{I}_{[1,p]}$. For example, one could take $\gamma_i(r) = \bar{\gamma}_i(r) - \bar{\gamma}_i(0)$, where $\bar{\gamma}_i(r) := \max_{E(x) \leq r} h_i(x, \kappa^f(x)) + r$. We can now specify the economic MPC scheme ensuring satisfaction of the relaxed transient average constraints (4.67) as follows.

Algorithm 4.55 (Economic MPC with relaxed transient average constraints). *Consider system* (2.1)*. At each time* $t \in \mathbb{I}_{\geq 0}$*, measure the state* $x(t)$*, solve Problem 4.43 with terminal region* $\mathbb{X}^f(t)$ *defined by* (4.46)–(4.47) *and* φ *given by*

$$\varphi(t) := \sum_{k=t-1}^{t-2+T} \gamma\Big(\alpha_0(1 - \frac{\lambda_{\min}(Q)}{\lambda_{\max}(P)})^k\Big) \tag{4.69}$$

for $t \in \mathbb{I}_{\geq 1}$ *and arbitrary* $\varphi(-N+1), \ldots, \varphi(0) \in \mathbb{R}_{\geq 0}$*, and apply the control input* $u(t) := u^0(0|t)$*.*

Theorem 4.56. *Let* $x_0 \in \mathbb{X}_N(0)$*, and suppose that Assumptions 2.1, 2.3, 2.6, 2.7 and 4.6 with continuous* κ^f *are satisfied. Then Problem 4.43 with terminal region* $\mathbb{X}^f(t)$ *and* φ *as specified in Algorithm 4.55 is feasible for all* $t \in \mathbb{I}_{\geq 0}$*, and the closed-loop system* (2.6) *resulting from application of Algorithm 4.55 satisfies the pointwise-in-time constraints* (4.40)*, the asymptotic average performance bound* (4.42) *and the (relaxed) transient average constraints* (4.67) *with*

$$\psi(t) = (1/T)\varphi(t - N + 1). \tag{4.70}$$

Proof. For proving recursive feasibility, the difference to Theorem 4.49 is the definition of $\mathbb{X}^f(t)$ and φ, i.e., we have to show that the candidate input and state sequences $\tilde{u}(t+1)$ and $\tilde{x}(t+1)$ as specified in the proof of Theorem 4.49 satisfy the constraints (4.62d)–(4.62f) with $\mathbb{X}^f(t)$ and φ as defined in Algorithm 4.55. Using Proposition 4.32, we conclude that $\tilde{x}(N|t+1) \in \mathbb{X}^f(t+1)$, i.e., the terminal constraint (4.62d) is satisfied. Feasibility of the constraints (4.62e) for all i and (4.62f) for $j \in \mathbb{I}_{[1,N-2]}$ can be established in the same way as in the proof of Theorem 4.49. Finally, for constraint (4.62f) with $j = N - 1$, consider the following. Using again Proposition 4.32, we conclude that for each $t \in \mathbb{I}_{\geq 0}$ and each $x \in \mathbb{X}_f(t)$, we have that $f(x, \kappa^f(x)) \in \mathbb{X}_f(t+1)$. Thus, due to the fact that $\tilde{x}(N-1|t+1) = x^0(N|t) \in \mathbb{X}_f(t)$, by (4.62a) and (4.62g) we obtain that $\tilde{x}(N+k|t+1) \in \mathbb{X}_f(t+k+1)$ for all $k \in \mathbb{I}_{[-1,T-2]}$. Furthermore, the definition of $\alpha_{\max}$ and $\mathbb{X}^f(t)$ in (4.46)–(4.47) implies that for each $t \in \mathbb{I}_{\geq 0}$, the terminal region $\mathbb{X}_f(t)$ is such that either $\alpha(t) = \alpha_0(1 - \frac{\lambda_{\min}(Q)}{\lambda_{\max}(P)})^t$ or $h(x, \kappa^f(x)) \in \mathbb{Y}$ (i.e., $h(x, \kappa^f(x)) \leq 0$) for all $x \in \mathbb{X}_f(t)$. Combining the above, we obtain

$$\sum_{k=N-1}^{N+T-2} h(\tilde{x}(k|t+1), \tilde{u}(k|t+1)) \leq \sum_{k=N-1}^{N+T-2} \gamma\Big(\alpha_0(1 - \frac{\lambda_{\min}(Q)}{\lambda_{\max}(P)})^{t+k-N+1}\Big) = \varphi(t+1),$$

where the first inequality follows from (4.68) (by considering that $\mathbb{X}_f(t) \subseteq \mathbb{X}_f(0)$ for all $t \in \mathbb{I}_{\geq 0}$ and $\tilde{x}(N+k|t+1) \in \mathbb{X}_f(t+k+1)$ for all $k \in \mathbb{I}_{[-1,T-2]}$ as established above), and the second equality follows by definition of $\varphi(t)$ in (4.69). This means that constraint (4.62f) with $j = N - 1$ is also satisfied at time $t + 1$, and hence recursive feasibility is established.

Fulfillment of the pointwise-in-time constraints (4.40) and the asymptotic average performance bound (4.42) for the closed-loop system (2.6) resulting from application of Algorithm 4.55 follow as in Theorem 4.49. Finally, fulfillment of the (relaxed) transient average constraints (4.67) for all $t \in \mathbb{I}_{\geq 0}$ follows directly from the first constraint of (4.62e) (i.e., with $i = 1$) and the definition of the receding horizon control law. Namely, from (4.62e) with $i = 1$ (this constraint applies for all $t \geq T - 1$) and the definition of the receding horizon control law, we obtain that along the closed-loop solution the following is satisfied for all $t \geq T - 1$:

$$\sum_{k=t-T+1}^{t} h(x(k), u(k)) \leq \varphi(t - T - N + 2).$$

Dividing by $T > 0$ and applying the index shift $t' = t - T + 1$, this is equivalent to the fact that (4.67) is satisfied for all $t \in \mathbb{I}_{\geq 0}$ with $\psi(t)$ given by (4.70). $\square$

In the following, we discuss various properties of Theorem 4.56 and the choice of the involved parameters.

Remark 4.57. *In order to obtain the tightest results via (4.70), one might want to choose small initial values $\varphi(-N + 1), \ldots, \varphi(0)$. On the other hand, they have to be large enough such that the constraints (4.62e)–(4.62f) are initially feasible. Furthermore, ψ given by (4.70) is nonincreasing if one chooses $\varphi(-N + 1) \geq \cdots \geq \varphi(0) \geq \varphi(1)$, and we have $\lim_{t\to\infty} \psi(t) = 0$ as required.*

Remark 4.58. *From (4.46)–(4.47), it follows that if $\alpha_{\max} > 0$, there exists some finite time $t' \in \mathbb{I}_{\geq 0}$ such that $\mathbb{X}^f(t) = \overline{\mathbb{X}}^f := \{x \in \mathbb{R}^n : E(x) \leq \min\{\alpha_0, \alpha_{\max}\}\}$ for all $t \in \mathbb{I}_{\geq t'}$, which by definition of $\alpha_{\max}$ implies that $h(x, \kappa^f(x)) \leq 0$ for all $x \in \overline{\mathbb{X}}^f$, i.e., Assumption 4.11 is satisfied for $\overline{\mathbb{X}}^f$. Hence for all $t \in \mathbb{I}_{\geq t'+1}$, one can define $\varphi(t) = 0$ instead of (4.69) and the results of Theorem 4.56 still hold. Using (4.70), this means that the original transient average constraints (4.60) are satisfied for all $t \in \mathbb{I}_{\geq t'+N}$.*

Remark 4.59. *In Theorem 4.56, relaxing Assumption 2.6 to Assumption 2.2, i.e., allowing for an unbounded constraint set $\mathbb{Z}$, can again be done as discussed in the paragraph below Theorem 2.5; again, note that these modifications are only necessary for the average performance bound (4.42).*

4.2.4 Numerical Examples

We now illustrate the concept of transient and asymptotic average constraints with two numerical examples. The first is an academic one, while the second considers a continuous flow stirred-tank reactor with parallel reactions.

Example 4.60. *Consider the two-dimensional system*

$$x(t + 1) = (1 - u(t)) \begin{bmatrix} \cos(\varepsilon + u(t)) & \sin(\varepsilon + u(t)) \\ -\sin(\varepsilon + u(t)) & \cos(\varepsilon + u(t)) \end{bmatrix} x(t), \tag{4.71}$$

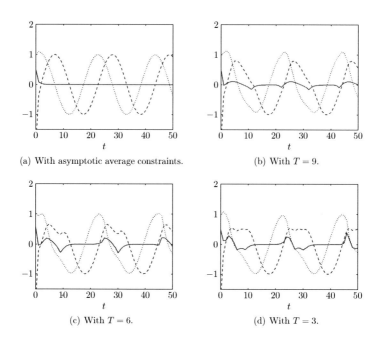

(a) With asymptotic average constraints.   (b) With $T = 9$.

(c) With $T = 6$.   (d) With $T = 3$.

Figure 4.4: Closed-loop state and input sequences in Example 4.60 with different values of T. Solid lines: input u; dashed lines: state x_1; dotted lines: state x_2.

where $x = \begin{bmatrix} x_1 & x_2 \end{bmatrix}^T \in \mathbb{R}^2$, $u \in \mathbb{R}$ and $0 < \varepsilon \leq \pi/2$. *The state and input constraint set is given by* $\mathbb{Z} = [-5,5]^2 \times [-2,2]$, *and the stage cost function is* $\ell(x,u) = (|x|-1)^2 + u^2$. *Note that the stage cost ℓ is positive definite with respect to the set* $\{(x,u) : |x| = 1, u = 0\}$; *furthermore, the unit sphere* $S_1 := \{x : |x| = 1\}$ *is invariant for system (4.71) under the constant control $u \equiv 0$ (in this case, the solution of (4.71) is a clockwise rotation whose angle of rotation per time step is determined by ε). Hence the optimal asymptotic average performance is obtained if the closed-loop system asymptotically converges to S_1 with $u \to 0$.*

We now consider the cases where asymptotic and transient average constraints of the form (2.14) and (4.60), respectively, are imposed on system (4.71), with $h(x,u) = x_1 - 0.5$ and $\mathbb{Y} = \mathbb{R}_{\leq 0}$. The sets of feasible and optimal state/input equilibrium pairs as defined in (2.15) and (2.16), respectively, are given by $S_{av} = \{(x,u) : x = 0, u \in [-2,2]\}$ and $S_{av}^* = \{(0,0)\}$. Figure 4.4 shows simulation results with $\varepsilon = 0.3$, prediction horizon $N = 10$ and initial condition $x_0 = \begin{bmatrix} -2 & 1 \end{bmatrix}^T$. The terminal cost $V^f(x) = x^2$, the local auxiliary control law $\kappa^f(x) = |x|$ and the terminal region $\mathbb{X}^f = \{x : |x| \leq 0.5\}$ were calculated such that Assumptions 4.3 (with constant terminal region $\mathbb{X}^f$) and 4.11 are satisfied; hence, as discussed in Section 4.2.1, also Assumptions 4.4–4.5 are, e.g., satisfied with $\overline{\mathbb{Y}} = \{0\}$ and $\sigma(t) = 0$ for all $t \in \mathbb{I}_{\geq 0}$. Figure 4.4(a) shows closed-loop sequences resulting from application of Algorithm 4.28, i.e., with asymptotic average constraints. One can see that the *resulting closed-loop system converges to B_1 with $u \to 0$, i.e., the same solution as without average constraints is obtained, as it satisfies the given asymptotic average constraint. The same behavior is obtained if transient average constraints with long enough period T are applied. On the other hand, if the period T is small, then converging to B_1 with $u \to 0$ becomes infeasible. Figures 4.4(b)–4.4(d) show closed-loop sequences resulting from application of Algorithm 4.47 with transient average constraint periods $T = 9$, $T = 6$ and $T = 3$, respectively. One can see that still some periodic orbit is obtained, but, loosely speaking, this periodic orbit is now shifted and squeezed together into the direction of smaller x_1 values. This effect increases with decreasing T, as the transient average constraints become more restrictive for smaller T. For smaller values of ε, this phenomenon can be encountered also for larger values of T, as the solution with $u = 0$ becomes "slower" and hence does not fulfill the transient average constraints anymore also for larger T.*

Example 4.61. *As a second example, a nonlinear continuous flow stirred-tank reactor is considered with two parallel reactions $R \to P_1$ and $R \to P_2$, where R is the reactant, P_1 is the desired product and P_2 is the waste product (Bailey et al., 1971). This system has also been considered previously in the context of economic MPC in (Angeli et al., 2011). We assume that the first reaction $R \to P_1$ is of second order, whereas the second one $R \to P_2$ is of first order. From the mass and energy balances, the following system equations can be derived (see (Bailey et al., 1971, Appendix I) for details):*

$$\dot{x}_1 = 1 - ax_1^2 e^{-1/x_3} - bx_1 e^{-c/x_3} - x_1$$
$$\dot{x}_2 = ax_1^2 e^{-1/x_3} - x_2 \qquad (4.72)$$
$$\dot{x}_3 = u - x_3$$

Herein, $x_1 = c_R/c_{R,in}$ is the dimensionless concentration of R inside the reactor (c_R and $c_{R,in}$ denote its concentration in the reactor and in the inlet stream, respectively), $x_2 = c_{P_1}/c_{R,in}$ is the dimensionless concentration of the desired product P_1 inside the reactor (c_{P_1} is its concentration in the reactor), and x_3 is the dimensionless temperature of the mixture in the reactor. The control input u is defined as $u = d + eQ$, where Q is the heat flux through the reactor wall, and a, b, c, d, e are constants depending on various process parameters, see (Bailey et al., 1971, Appendix I). The control objective is to maximize the amount of product P_1, which translates into a stage cost function $\ell(x, u) = -x_2$. For our simulations, we use the same parameter values as in (Bailey et al., 1971), i.e., $a = 10^4$, $b = 400$, $c = 0.55$, and state and input constraint set $\mathbb{Z} = \mathbb{R}_{\geq 0}^3 \times [0.049, 0.449]$. With these parameter values, one obtains $x^* = \begin{bmatrix} 0.0832 & 0.0846 & 0.1491 \end{bmatrix}^T$ and $u^* = 0.1491$; furthermore, as was shown in (Bailey et al., 1971), periodic solutions can outperform steady-state operation of the system.

The system is discretized with a sample time $T_s = 1/10$, and a prediction horizon of $N = 50$ is chosen; the simulations were implemented in Matlab using integrators from the ACADO toolkit (Houska et al., 2011). The terminal cost, the terminal region, and the local auxiliary control law were calculated according to the procedure in (Amrit et al., 2011) such that Assumptions 4.3 (with constant terminal region $\mathbb{X}^f$) and 4.11 are satisfied; hence, as discussed in Section 4.2.1, also Assumptions 4.4–4.5 are, e.g., satisfied with $\overline{\mathbb{Y}} = \{0\}$ and $\sigma(t) = 0$ for all $t \in \mathbb{I}_{\geq 0}$. Figure 4.5 (left column) shows simulation results for the input u and the concentration of the desired product x_2 resulting from application of Algorithm 2.3, i.e., without average constraints; as expected, the closed-loop system

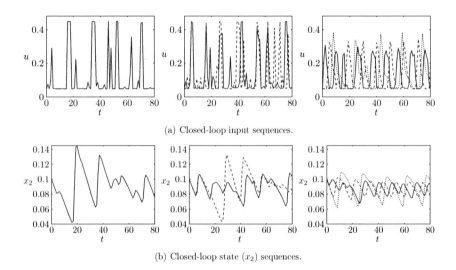

(a) Closed-loop input sequences.

(b) Closed-loop state (x_2) sequences.

Figure 4.5: Closed-loop input and state sequences in Example 4.61. Left: without average constraints; middle: with asymptotic average constraints and $y_{00} = 0.5$ (solid) and $y_{00} = 1$ (dashed); right: with transient average constraints and $T = 5$ (solid), $T = 10$ (dashed) and $T = 15$ (dotted).

is not convergent. We now consider the cases where asymptotic and transient average constraints of the form (2.14) and (4.60), respectively, are imposed on system (4.72), with $h(x, u) = (u - u^*)^2 - 0.01$ and $\mathbb{Y} = \mathbb{R}_{\leq 0}$. This can be interpreted as that the averaged deviation of the heat flux Q from its value at the optimal steady-state has to be bounded by some constant. Figure 4.5 (middle and right column) shows closed-loop input and state sequences resulting from application of Algorithms 4.28 and 4.47, respectively. For the simulations with asymptotic average constraints, in (4.38) we used $\mathbb{Y}_{00} = \{y \in \mathbb{R} : |y| \leq y_{00}\}$ with different values of y_{00}. One can see that as discussed in Remark 4.25, a larger value of y_{00} is less restrictive, resulting in more deviation of the closed-loop input and state sequences from u^* and x_2^*, respectively. The simulations with transient average constraints reveal that as expected, smaller values of T are more restrictive, resulting again in less deviation of the closed-loop input and state sequences from u^* and x_2^*, respectively. This can also be seen when evaluating the closed-loop average performance, i.e., computing $J = \sum_{t=0}^{t_{end}} -x_2(t)/(t_{end}+1)$, where t_{end} is the simulation end time. Without average constraints, we obtain $J = -0.0909$, with asymptotic average constraints we have $J = -0.0889$ for $y_{00} = 1$ and $J = -0.0885$ for $y_{00} = 0.5$, and with transient average constraints we obtain $J = -0.0878$ for $T = 15$, $J = -0.0874$ for $T = 10$ and $J = -0.0867$ for $T = 5$. Note that as expected, all these values are better than the optimal steady-state cost, which is given as $\ell(x^*, u^*) = -x_2^* = -0.0846$.

4.3 Economic MPC with self-tuning terminal cost

In the previous sections of this chapter, we have considered economic MPC schemes where the terminal predicted state was required to be equal to the optimal steady-state x^* (respectively, had to be contained in some terminal region around x^*). In this section, we propose an economic MPC setting involving a *generalized* terminal state constraint, meaning that the terminal predicted state has to be equal to some arbitrary steady-state instead of x^*. In the literature, such a setup has first been proposed in the context of stabilizing tracking MPC by (Ferramosca et al., 2009; Limon et al., 2008) and (Fagiano and Teel, 2012, 2013), and recently also within an economic MPC framework (Fagiano and Teel, 2012, 2013; Ferramosca, 2011; Ferramosca et al., 2010). The big advantage of such a generalized terminal constraint is that the set $\mathbb{X}_N$ of feasible states is possibly much larger than with a fixed terminal constraint, and furthermore a loss of feasibility can be prevented which otherwise might occur if the cost function (and hence the optimal steady-state) changes online. In particular, in (Ferramosca, 2011; Ferramosca et al., 2010), the authors propose to use a slightly modified economic cost function and in addition an offset cost function which penalizes the distance of the predicted terminal steady-state $x(N|t)$ to the optimal steady-state x^*. On the other hand, in (Fagiano and Teel, 2012, 2013) the optimal steady-state x^* does not have to be known a priori, and the authors use the original (economic) cost function and in addition a terminal cost term which penalizes the economic cost of the predicted terminal steady-state $x(N|t)$. It is shown that if the terminal weight is large enough, then the cost of the predicted terminal steady-state will be arbitrarily close to the cost of the best reachable steady-state. Furthermore, under additional assumptions and by overriding the MPC algorithm, i.e., if necessary, following the previously optimal solution, the cost of the predicted terminal steady-state, and hence also the average performance of the closed-loop system, will eventually be arbitrarily close to the cost of the best overall steady-state (Fagiano and Teel, 2013).

In the following, we propose an economic MPC scheme with self-tuning terminal cost, which is based on the scheme presented in (Fagiano and Teel, 2013). The advantage of such a self-tuning terminal cost in comparison to a fixed terminal cost is that we do not necessarily need to make the terminal weight large in order to guarantee certain performance properties. This is desirable both from a numerical point of view as well as in order not to modify the original economic cost function too much. In fact, we illustrate with a simple example that in some cases, a smaller terminal weight leads to a better closed-loop average performance than a large terminal weight. Furthermore, this allows us to establish (average) performance guarantees for the closed-loop system without possibly overriding the MPC algorithm as was done in (Fagiano and Teel, 2013). For clarity of presentation, we develop our results using a generalized terminal equality constraint, before discussing extensions to a generalized terminal region setting (see Section 4.3.4).

4.3.1 Problem setup

Consider again the economic MPC setting of Chapter 2, namely system (2.1) subject to the pointwise-in-time state and input constraints (2.2), i.e., $(x(t), u(t)) \in \mathbb{Z} \subseteq \mathbb{X} \times \mathbb{U}$ for all $t \in \mathbb{I}_{\geq 0}$. Furthermore, recall the definition of the set of feasible and optimal state/input equilibrium pairs S and S^* from (2.3) and (2.7), respectively. We assume without loss of generality that $\ell(x^*, u^*) = 0$. Before stating the proposed MPC scheme with self-tuning terminal cost, we need the following definitions. The set of steady-states which are reachable in $N \in \mathbb{I}_{\geq 1}$ steps from a point $y \in \mathbb{X}$ is denoted by

$$
\mathcal{R}_N^s(y) := \{ x \in \mathbb{X} : \; \exists \boldsymbol{v} : \mathbb{I}_{[0,N]} \to \mathbb{U} \text{ s.t. } z(0) = y, \; z(t+1) = f(z(t), v(t)) \; \forall t \in \mathbb{I}_{[0,N-1]},
$$
$$
z(N) = x, \; (x, v(N)) \in S, \; (z(t), v(t)) \in Z \; \forall t \in \mathbb{I}_{[0,N]} \}. \tag{4.73}
$$

Note that for each $y \in \mathbb{X}$, the set $\mathcal{R}_N^s(y)$ is compact in case that Assumptions 2.1 and 2.2 are satisfied (as then the N-step reachable set from a point y using controls in a compact set $\mathbb{U}$ is compact). Define the best achievable steady-state cost from a point $y \in \mathbb{X}$ as[17]

$$
\ell_{\min}(y) := \min_{x,u} \ell(x, u)
$$
$$
\text{s.t. } x \in \mathcal{R}_N^s(y), \; (x, u) \in Z, \; x = f(x, u). \tag{4.74}
$$

Furthermore, in the following we need the notion of the best *robustly* achievable steady-state cost from a point $y \in \mathbb{X}$, which we define as follows. For each $\varepsilon \geq 0$, denote by

$$
\ell_{\min}'(y; \varepsilon) := \sup_{z \in B_\varepsilon(y) \cap \mathbb{X}} \ell_{\min}(z) \tag{4.75}
$$

the supremum of the best achievable steady-state cost on the set $B_\varepsilon(y) \cap \mathbb{X}$. With this, we define the best robustly achievable steady-state cost from a point $y \in \mathbb{X}$ as

$$
\overline{\ell}_{\min}(y) := \lim_{\varepsilon \searrow 0} \ell_{\min}'(y; \varepsilon). \tag{4.76}
$$

Note that the limit in (4.76) exists as $\ell_{\min}'(y; \varepsilon)$ is monotonically nonincreasing when ε decreases to zero. From the definitions in (4.74) and (4.76), it immediately follows that for each $y \in \mathbb{X}$ we have $\ell_{\min}(y) \leq \overline{\ell}_{\min}(y)$; however, equality does in general not hold as $\ell_{\min}'(y; \varepsilon)$ is not necessarily continuous in ε at $\varepsilon = 0$ (for a simple example of this fact, see Example 4.81 in Section 4.3.5).

In order to define the economic MPC scheme with self-tuning terminal cost, consider the following optimization problem at each time $t \in \mathbb{I}_{\geq 0}$ with measured state $x(t)$, which is a variation of the one proposed in (Fagiano and Teel, 2013) with fixed terminal weight.

Problem 4.62.

$$
\underset{\boldsymbol{u}(t)}{minimize} \; J_N(x(t), \boldsymbol{u}(t), \beta(t))
$$

subject to

$$
x(k+1|t) = f(x(k|t), u(k|t)), \quad k \in \mathbb{I}_{[0,N-1]} \tag{4.77a}
$$
$$
x(0|t) = x(t) \tag{4.77b}
$$
$$
(x(k|t), u(k|t)) \in Z, \quad k \in \mathbb{I}_{[0,N]} \tag{4.77c}
$$
$$
x(N|t) = f(x(N|t), u(N|t)), \tag{4.77d}
$$
$$
\ell(x(N|t), u(N|t)) \leq \kappa(t), \tag{4.77e}
$$

where

$$
J_N(x(t), \boldsymbol{u}(t), \beta(t)) := \sum_{k=0}^{N-1} \ell(x(k|t), u(k|t)) + \beta(t)\ell(x(N|t), u(N|t)). \tag{4.78}
$$

---

[17]In the following, if a minimum is taken over the empty set, then by convention the minimum is $+\infty$.

As pointed out earlier, the special feature of Problem 4.62 and its main difference to the standard economic MPC Problem 2.2 is the generalized terminal constraint (4.77d), meaning that the predicted terminal state $x(N|t)$ has to be equal to an arbitrary steady-state (satisfying (4.77e)) and not to a specific one. Furthermore, note that compared to Problem 2.2, the predicted input sequence $\boldsymbol{u}(t)$ in addition contains $u(N|t)$ (which appears in (4.77d) and (4.77e)), i.e., $\boldsymbol{u}(t) := \{u(0|t), \ldots, u(N|t)\}$. As before, denote the minimizer of Problem (4.62) by $\boldsymbol{u}^0(t) := \{u^0(0|t), \ldots, u^0(N|t)\}$, the corresponding state sequence by $\boldsymbol{x}^0(t) := \{x^0(0|t), \ldots, x^0(N|t)\}$, and the optimal value function by $J_N^0(x(t), \beta(t)) := J_N(x(t), \boldsymbol{u}^0(t), \beta(t))$, which depends on the terminal weight $\beta(t)$. Furthermore, as before, $\mathbb{X}_N(t)$ denotes the set of all states $x \in \mathbb{X}$ such that Problem 4.62 (with $x(t) = x$) has a solution. The parameter $\kappa(t)$ and the terminal weight $\beta(t)$ in (4.77e) and (4.78), respectively, are updated as follows:

$$\kappa(t+1) = \ell(x^0(N|t), u^0(N|t)), \qquad \kappa(0) = \kappa_0 \geq 0. \tag{4.79}$$
$$\beta(t+1) = B(\beta(t), x(t), \kappa(t+1)), \qquad \beta(0) = \beta_0 \geq 0, \tag{4.80}$$

In Section 4.3.3, we will specify and discuss several particular update rules B, and also further discuss the initialization of κ in Section 4.3.4. The economic MPC scheme with self-tuning terminal cost is now given as follows.

Algorithm 4.63 (Economic MPC with self-tuning terminal cost). *Consider system* (2.1). *At each time $t \in \mathbb{I}_{\geq 0}$, measure the state $x(t)$, solve Problem 4.62 with $\kappa(t)$ and $\beta(t)$ determined by (4.79)–(4.80) and apply the control input $u(t) := u^0(0|t)$.*

Note that (4.79) together with (4.77e) implies that at each time t, the cost of the terminal predicted state has to be less or equal than the cost of the terminal predicted state at time $t - 1$. This means that the closed-loop sequence $\boldsymbol{\kappa} := \{\kappa(0), \kappa(1), \ldots\}$ resulting from application of Algorithm 4.63 is nonincreasing and bounded from below (by $0 = \ell(x^*, u^*)$) and hence it converges. Denote the limit by $\kappa_\infty := \lim_{t \to \infty} \kappa(t) \geq 0$. Note that the sequence $\boldsymbol{\kappa}$ is convergent irrespective of the evolution of the terminal weight β, however, the limit κ_∞ does in general depend on β. As noted above, in the following we want to analyze the behavior of the closed-loop system (2.6) resulting from application of Algorithm 4.63 when using a self-tuning, time-varying terminal weight β which is not unnecessarily large, and without further modifying the MPC algorithm as in (Fagiano and Teel, 2013, Algorithm 3). Before doing so, we briefly motivate with a simple example that larger values of β do not necessarily lead to smaller values of κ_∞ and to a better average performance of the closed-loop system.

Example 4.64. *Consider the scalar system $x(t + 1) = x(t)u(t)$ with state and input constraint set $\mathbb{Z} = \mathbb{X} \times \mathbb{U}$ with $\mathbb{X} = [-5, 5]$ and $\mathbb{U} = [-1.2, 1.2]$, cost function*

$$\ell(x, u) = \frac{1}{4}x^4 - \frac{4}{3}x^3 + \frac{3}{2}x^2 + \frac{9}{4} + (u - 1)^2$$

and prediction horizon $N = 1$. The cost function ℓ is plotted in Figure 4.6(a) for $u = 1$. Figure 4.6(b) shows closed-loop state sequences resulting from application of Algorithm 4.63 with $x_0 = 1.1$, $\kappa_0 = \max_{(x,u) \in \mathbb{Z}} \ell(x, u)$ and four different constant values of β. As can be seen, x converges to 0 and hence $\kappa_\infty = \ell(0, 1) = 9/4$ for both $\beta \equiv 1.5$ and $\beta \equiv 5$ (the same also happens for all larger values of β), whereas x converges to 3 and hence $\kappa_\infty = \ell(3, 1) = 0$

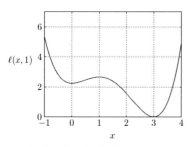

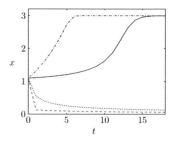

(a) Cost function ℓ with $u = 1$.

(b) State sequences with $\beta \equiv 0.1$ (solid), $\beta \equiv 1$ (dash-dotted), $\beta \equiv 1.5$ (dotted), and $\beta \equiv 5$ (dashed).

Figure 4.6: Cost function and closed-loop state sequences in Example 4.64.

for both $\beta \equiv 0.1$ and $\beta \equiv 1$. The reason for this is that for sufficiently large values of β, the control action is chosen such that the cost of the predicted terminal steady-state $x(1|t)$ becomes small. As can be seen from Figure 4.6(a), the best steady-state which can be reached within one step (given the above input constraints) is $x = 0$, and hence u is chosen such that x decreases. On the other hand, for small values of β, the terminal steady-state is not weighted as much, which results in an increasing x and a better average performance of the resulting closed-loop system.

4.3.2 Average performance with time-varying terminal weight

In this section, we study the asymptotic average performance of the closed-loop system (2.6) resulting from application of Algorithm 4.63. To this end, we consider the following two assumptions on the closed-loop terminal weight sequence $\boldsymbol{\beta} := \{\beta(0), \beta(1), \dots\}$ resulting from application of Algorithm 4.63. Later, in Section 4.3.3, we propose several update rules B for the terminal weight such that these assumptions are satisfied. Now let $\gamma(t) := \beta(t+1) - \beta(t)$.

Assumption 4.12. *The closed-loop sequence $\boldsymbol{\beta}$ as determined by (4.80) and resulting from application of Algorithm 4.63 satisfies $\beta(t) \geq 0$ and $\gamma(t) \leq c$ for all $t \in \mathbb{I}_{\geq 0}$ and some constant $c \in \mathbb{R}$, and $\limsup_{t \to \infty} \gamma(t) \leq 0$.*

Assumption 4.13. *The closed-loop sequence $\boldsymbol{\beta}$ as determined by (4.80) and resulting from application of Algorithm 4.63 satisfies $\beta(t) \geq 0$ and $\gamma(t) \leq c$ for all $t \in \mathbb{I}_{\geq 0}$ and some constant $c \in \mathbb{R}$, and $\liminf_{t \to \infty} \beta(t) < \infty$.*

Theorem 4.65. *Let $x_0 \in \mathbb{X}_N(0)$, and suppose that Assumptions 2.1, 2.3 and 2.6 are satisfied. Then Problem 4.62 with $\kappa(t)$ and $\beta(t)$ determined by (4.79)–(4.80) is feasible for all $t \in \mathbb{I}_{\geq 0}$, and the closed-loop system (2.6) resulting from application of Algorithm 4.63 satisfies the pointwise-in-time state and input constraints (4.40) as well as the following.*

If Assumption 4.12 is satisfied, then

$$\limsup_{T \to \infty} \frac{\sum_{t=0}^{T-1} \ell(x(t), u(t))}{T} \leq \kappa_\infty. \tag{4.81}$$

If Assumption 4.13 is satisfied, then

$$\liminf_{T\to\infty} \frac{\sum_{t=0}^{T-1} \ell(x(t), u(t))}{T} \le \kappa_\infty. \tag{4.82}$$

Proof. Recursive feasibility of Problem 4.62 can again be established by considering that $\tilde{\boldsymbol{u}}(t+1) := \{u^0(1|t), \ldots, u^0(N|t), u^0(N|t)\}$ is a feasible solution at time $t+1$ given that Problem 4.62 was feasible at time t, and satisfaction of the pointwise-in-time state and input constraints (4.40) for the closed-loop system results again from the definition of the input $u(t)$ in Algorithm 4.63. To establish the average performance bounds (4.81) and (4.82), consider the following. Using the candidate input sequence $\tilde{\boldsymbol{u}}(t+1)$, we obtain

$$\begin{aligned}
J_N^0(x(t+1), &\beta(t+1)) - J_N^0(x(t), \beta(t)) \\
&\le J(x(t+1), \tilde{\boldsymbol{u}}(t+1), \beta(t+1)) - J(x(t), \boldsymbol{u}^0(t), \beta(t)) \\
&= (1 + \gamma(t))\ell(x^0(N|t), u^0(N|t)) - \ell(x(t), u(t)).
\end{aligned} \tag{4.83}$$

As discussed above, the sequence $\ell(x^0(N|t), u^0(N|t))$ is non-increasing in t and converges to κ_∞ for $t \to \infty$. This means that $\varepsilon(t) := \ell(x^0(N|t), u^0(N|t)) - \kappa_\infty$ converges to zero for $t \to \infty$. Furthermore, we have $0 \le \varepsilon(t) \le \varepsilon(0)$ for all $t \in \mathbb{I}_{\ge 0}$. Summing the inequality in (4.83), for each $T \ge 1$ we obtain

$$J_N^0(x(T), \beta(T)) - J_N^0(x_0, \beta_0) \le \sum_{t=0}^{T-1} \Big((1 + \gamma(t))(\kappa_\infty + \varepsilon(t)) - \ell(x(t), u(t))\Big) \tag{4.84}$$

Now first consider the case where Assumption 4.12 is satisfied. Taking averages on both sides of (4.84), we obtain

$$\liminf_{T\to\infty}(1/T)\big(J_N^0(x(T), \beta(T)) - J_N^0(x_0, \beta_0)\big)$$

$$\le \liminf_{T\to\infty}(1/T)\Big(T\kappa_\infty + \sum_{t=0}^{T-1}\big(\gamma(t)\kappa_\infty + (1 + \gamma(t))\varepsilon(t) - \ell(x(t), u(t))\big)\Big)$$

$$\le \kappa_\infty + \liminf_{T\to\infty}(1/T)\Big(\sum_{t=0}^{T-1} -\ell(x(t), u(t))\Big) + \limsup_{T\to\infty}(1/T)\Big(\sum_{t=0}^{T-1}\big(\gamma(t)\kappa_\infty + (1 + \gamma(t))\varepsilon(t)\big)\Big)$$

$$\le \kappa_\infty - \limsup_{T\to\infty}(1/T)\Big(\sum_{t=0}^{T-1}\ell(x(t), u(t))\Big), \tag{4.85}$$

where the last inequality follows from the fact that $\lim_{t\to\infty} \varepsilon(t) = 0$, $\limsup_{t\to\infty}\gamma(t) \le 0$, $\kappa_\infty \ge 0$, and $0 \le \varepsilon(t) \le \varepsilon(0) < \infty$ and $\gamma(t) \le c < \infty$ for all $t \in \mathbb{I}_{\ge 0}$. On the other hand, as $\beta(t) \ge 0$ for all $t \in \mathbb{I}_{\ge 0}$ and furthermore ℓ is continuous, $\mathbb{Z}$ compact, and $\ell(x(N|t), u(N|t)) \ge \ell(x^*, u^*) = 0$, it follows by definition of the cost function J_N in (4.78) that there exists a constant $\underline{J} \in \mathbb{R}$ such that $J_N^0(x(t), \beta(t)) \ge \underline{J}$ for all $t \in \mathbb{I}_{\ge 0}$. Hence we obtain

$$\liminf_{T\to\infty} \frac{J_N^0(x(T), \beta(T)) - J_N^0(x_0, \beta_0)}{T} \ge \liminf_{T\to\infty} \frac{J_N^0(x(T), \beta(T))}{T} + \liminf_{T\to\infty} \frac{-J_N^0(x_0, \beta_0)}{T}$$

$$\ge \liminf_{T\to\infty}(\underline{J}/T) + \liminf_{T\to\infty} \frac{-J_N^0(x_0, \beta_0)}{T} = 0. \tag{4.86}$$

Combining (4.85) and (4.86) yields that (4.81) is satisfied, which concludes the proof of the first statement of Theorem 4.65.

Second, consider the case where Assumption 4.13 is satisfied. Then, there exists an infinite sequence of time instants $\{t_i\} \subseteq \mathbb{I}_{\geq 0}$ such that $\beta(t_i) \leq \beta_\infty$ for all t_i and some $\beta_\infty < \infty$. Using the fact that $\sum_{t=0}^{t_i-1} \gamma(t) = \beta(t_i) - \beta_0$ by definition of γ, and $\gamma(t) \leq c$ for all $t \in \mathbb{I}_{\geq 0}$ by assumption, from (4.84) we obtain that

$$J_N^0(x(t_i), \beta(t_i)) - J_N^0(x_0, \beta_0) \leq (t_i + \beta_\infty - \beta_0)\kappa_\infty + \sum_{t=0}^{t_i-1}\big((1+c)\varepsilon(t) - \ell(x(t), u(t))\big).$$

Taking averages on both sides, we have

$$\liminf_{T\to\infty}(1/T)\big(J_N^0(x(T), \beta(T)) - J_N^0(x_0, \beta_0)\big) \leq \liminf_{i\to\infty}(1/t_i)\big(J_N^0(x(t_i), \beta(t_i)) - J_N^0(x_0, \beta_0)\big)$$

$$\leq \liminf_{i\to\infty}(1/t_i)\Big((t_i + \beta_\infty - \beta_0)\kappa_\infty + \sum_{t=0}^{t_i-1}\big((1+c)\varepsilon(t) - \ell(x(t), u(t))\big)\Big)$$

$$\leq \kappa_\infty + \liminf_{i\to\infty}(1/t_i)\Big(\sum_{t=0}^{t_i-1}\big((1+c)\varepsilon(t) - \ell(x(t), u(t))\big)\Big)$$

$$\leq \kappa_\infty - \limsup_{i\to\infty}(1/t_i)\Big(\sum_{t=0}^{t_i-1}\ell(x(t), u(t))\Big) \leq \kappa_\infty - \liminf_{T\to\infty}(1/T)\Big(\sum_{t=0}^{T-1}\ell(x(t), u(t))\Big), \quad (4.87)$$

where the fourth inequality is again due to the fact that $\lim_{t\to\infty}\varepsilon(t) = 0$ and $\varepsilon(t) \leq \varepsilon(0) < \infty$ for all $t \in \mathbb{I}_{\geq 0}$, and the last inequality follows from the fact that $\{t_i\}$ is an infinite subsequence of $\mathbb{I}_{\geq 0}$. Combining (4.87) with (4.86) (which is valid independent of whether Assumption 4.12 or 4.13 is satisfied) results in (4.82), which concludes the proof of the second statement of Theorem 4.65. □

Remark 4.66. *Note that a constant β trivially satisfies both Assumptions 4.12 and 4.13. The motivation for studying more complex, time-varying terminal weights β is that in this case, statements about the value of κ_∞, and hence, by Theorem 4.65, also on the average performance of the closed-loop system, can be made (see Section 4.3.3).*

Remark 4.67. *Assumption 4.13 is weaker than requiring β to be bounded. In fact, if Assumption 4.13 is strengthened to β being bounded, then again the stronger conclusion (4.81) instead of (4.82) follows. Namely, if $\beta(t) \leq \beta_\infty$ for all $t \in \mathbb{I}_{\geq 0}$, the calculations in (4.87) hold for all $t \in \mathbb{I}_{\geq 0}$ and not only for the subsequence $\{t_i\} \subseteq \mathbb{I}_{\geq 0}$. Then, the expression in front of the last inequality in (4.87) together with (4.86) imply that (4.81) is satisfied.*

Remark 4.68. *In Theorem 4.65, relaxing Assumption 2.6 to Assumption 2.2, i.e., allowing for an unbounded constraint set $\mathbb{Z}$, can again be done similar to what was discussed in the paragraph below Theorem 2.5. Namely, one has to ensure that the left hand side of the first line in (4.86) is greater or equal to zero, which is, e.g., the case if ℓ is assumed to be bounded from below on $\mathbb{Z}$.*

4.3.3 Self-tuning terminal weights

In this section, we propose several self-tuning update rules B for the terminal weight β. Having established that κ_∞ is an upper bound for the average performance of the closed-loop system, we want to find conditions such that bounds for κ_∞ can be guaranteed. In this respect, we first formulate general conditions on update rules, before presenting specific update rules satisfying these conditions.

Specifications for update rules B

Let $\omega_B(x_0)$ be the ω-limit set of the closed-loop state sequence $\boldsymbol{x}$ resulting from application of Algorithm 4.63 and with initial condition x_0, i.e., $\omega_B(x_0) := \{y \in \mathbb{X} : \exists$ infinite sequence $\{t_n\} \subseteq \mathbb{I}_{\geq 0}$ s.t $x(0) = x_0$ and $\lim_{n \to \infty} x(t_n) = y\}$. Note that $\omega_B(x_0)$ is compact and non-empty if Assumption 2.6 is satisfied, i.e., $\mathbb{Z}$ is compact. We then have the following result:

Theorem 4.69. *Let $x_0 \in \mathbb{X}_N(0)$, and suppose that Assumptions 2.1, 2.3 and 2.6 are satisfied. We then have the following.*
(i) Suppose that the update rule B is such that for all sequences $\boldsymbol{x} : \mathbb{I}_{\geq 0} \to \mathbb{X}$ and $\boldsymbol{\kappa} : \mathbb{I}_{\geq 0} \to \mathbb{R}$ with $\boldsymbol{\kappa}$ nonincreasing and $\lim_{t \to \infty} \kappa(t) = \kappa_\infty$, regarded as open-loop input signals in (4.80), it holds that

$$\kappa_\infty - \liminf_{t \to \infty} \ell_{\min}(x(t)) > 0 \;\Rightarrow\; \liminf_{t \to \infty} \beta(t) = \infty. \tag{4.88}$$

Then, for the closed-loop system (2.6) resulting from application of Algorithm 4.63, it holds that $\lim_{t \to \infty} \ell_{\min}(x(t))$ exists and

$$\kappa_\infty = \lim_{t \to \infty} \ell_{\min}(x(t)) \leq \inf_{y \in \omega_B(x_0)} \overline{\ell}_{\min}(y). \tag{4.89}$$

(ii) Suppose that the update rule B is such that for all sequences $\boldsymbol{x} : \mathbb{I}_{\geq 0} \to \mathbb{X}$ and $\boldsymbol{\kappa} : \mathbb{I}_{\geq 0} \to \mathbb{R}$ with $\boldsymbol{\kappa}$ nonincreasing and $\lim_{t \to \infty} \kappa(t) = \kappa_\infty$, regarded as open-loop input signals in (4.80), it holds that

$$\kappa_\infty - \limsup_{t \to \infty} \ell_{\min}(x(t)) > 0 \;\Rightarrow\; \limsup_{t \to \infty} \beta(t) = \infty. \tag{4.90}$$

Then, for the closed-loop system (2.6) resulting from application of Algorithm 4.63, it holds that

$$\kappa_\infty = \limsup_{t \to \infty} \ell_{\min}(x(t)) \leq \sup_{y \in \omega_B(x_0)} \overline{\ell}_{\min}(y). \tag{4.91}$$

Remark 4.70. *Conditions (4.88) and (4.90) in Theorem 4.69 have to be understood for "open-loop" sequences $\boldsymbol{x}$, $\boldsymbol{\kappa}$ and $\boldsymbol{\beta}$, i.e., when some (given) sequences $\boldsymbol{x}$ and $\boldsymbol{\kappa}$ are regarded as open-loop input signals in the update rule B (4.80). The assertions (4.89) and (4.91), on the other hand, hold for "closed-loop" sequences $\boldsymbol{x}$, $\boldsymbol{\kappa}$ and $\boldsymbol{\beta}$ resulting from application of Algorithm 4.63.*

Proof of Theorem 4.69. (i) First, we show that $\kappa_\infty - \liminf_{t \to \infty} \ell_{\min}(x(t)) \leq 0$ holds for the closed-loop system (2.6) resulting from application of Algorithm 4.63 if (4.88) is satisfied. Namely, assume for contradiction that $\kappa_\infty - \liminf_{t \to \infty} \ell_{\min}(x(t)) > 0$ holds for the closed-loop system (2.6) resulting from application of Algorithm 4.63. We can then use the (closed-loop) sequences $\boldsymbol{x}$ and $\boldsymbol{\kappa}$ in (4.88) to conclude that the corresponding (closed-loop) terminal weight sequence $\boldsymbol{\beta}$ satisfies $\liminf_{t \to \infty} \beta(t) = \infty$. However, in (Fagiano and Teel, 2013, Proposition 2) it was shown that for each $\delta > 0$, if $\beta > a/\delta$ for some finite constant $a > 0$, then $\ell(x^0(N|t), u^0(N|t)) \leq \ell_{\min}(x(t)) + \delta$ for all $x(t)$, which results from Assumptions 2.1, 2.3 and 2.6, i.e., continuity of f and ℓ and compactness of $\mathbb{Z}$. But then, as $\liminf_{t\to\infty} \beta(t) = \infty$, we obtain that $\kappa_\infty \leq \liminf_{t\to\infty} \ell_{\min}(x(t))$, which contradicts our assumption. Thus it holds that $\kappa_\infty - \liminf_{t\to\infty} \ell_{\min}(x(t)) \leq 0$, which by the definition of κ_∞ and $\ell_{\min}$ implies that $\lim_{t\to\infty} \ell_{\min}(x(t))$ exists and $\kappa_\infty - \lim_{t\to\infty} \ell_{\min}(x(t)) = 0$. This establishes the equality in (4.89). Second, according to the definition of the ω-limit set $\omega_B(x_0)$, for each $y \in \omega_B(x_0)$ and each $\varepsilon > 0$ there exists an infinite sequence of time instants $\{t_i^y\} \subseteq \mathbb{I}_{\geq 0}$ such that $x(t_i^y) \in B_\varepsilon(y) \cap \mathbb{Z}_{\mathbb{X}}$. But this implies that

$$\kappa_\infty = \lim_{t\to\infty} \ell_{\min}(x(t)) = \lim_{i\to\infty} \ell_{\min}(x(t_i^y)) \leq \ell'_{\min}(y;\varepsilon).$$

As this holds for each $\varepsilon > 0$ and each $y \in \omega_B(x_0)$, we obtain

$$\kappa_\infty = \lim_{t\to\infty} \ell_{\min}(x(t)) \leq \inf_{y\in\omega_B(x_0)} \bar{\ell}_{\min}(y),$$

which establishes statement (i) of the theorem.

(ii) The proof of this statement is similar the proof of statement (i). We first show that $\kappa_\infty - \limsup_{t\to\infty} \ell_{\min}(x(t)) \leq 0$ for the closed-loop system (2.6) resulting from application of Algorithm 4.63 if (4.90) is satisfied. Namely, assume for contradiction that $\kappa_\infty - \limsup_{t\to\infty} \ell_{\min}(x(t)) > 0$ for the closed-loop system (2.6) resulting from application of Algorithm 4.63. We can then use the (closed-loop) sequences $\boldsymbol{x}$ and $\boldsymbol{\kappa}$ in (4.90) to conclude that the corresponding (closed-loop) terminal weight sequence $\boldsymbol{\beta}$ satisfies $\limsup_{t\to\infty} \beta(t) = \infty$. As above, we can now use the result from (Fagiano and Teel, 2013, Proposition 2) that for each $\delta > 0$, if $\beta > a/\delta$ for some finite constant $a > 0$, then $\ell(x^0(N|t), u^0(N|t)) \leq \ell_{\min}(x(t)) + \delta$ for all $x(t)$. Namely, as $\limsup_{t\to\infty} \beta(t) = \infty$, we then obtain that $\kappa_\infty \leq \limsup_{t\to\infty} \ell_{\min}(x(t))$, which contradicts our assumption. By definition of κ_∞ and $\ell_{\min}$, the above inequality has to hold with equality, i.e., $\kappa_\infty = \limsup_{t\to\infty} \ell_{\min}(x(t))$, which establishes the equality in (4.91). Now let $\{t_i\} \subseteq \mathbb{I}_{\geq 0}$ be an infinite sequence of time instants such that $\lim_{i\to\infty} \ell_{\min}(x(t_i))$ exists and satisfies $\lim_{i\to\infty} \ell_{\min}(x(t_i)) = \limsup_{t\to\infty} \ell_{\min}(x(t))$. According to the definition of the ω-limit set $\omega_B(x_0)$, there exists a point $y^* \in \omega_B(x_0)$ and for each $\varepsilon > 0$ an infinite subsequence $\{t_r\}$ of the sequence $\{t_i\}$ such that $x(t_r) \in B_\varepsilon(y^*) \cap \mathbb{Z}_{\mathbb{X}}$. But this implies that

$$\limsup_{t\to\infty} \ell_{\min}(x(t)) = \lim_{r\to\infty} \ell_{\min}(x(t_r)) \leq \ell'_{\min}(y^*;\varepsilon).$$

But as this holds for each $\varepsilon > 0$, we obtain

$$\kappa_\infty = \limsup_{t\to\infty} \ell_{\min}(x(t)) \leq \bar{\ell}_{\min}(y^*) \leq \sup_{y\in\omega_B(x_0)} \bar{\ell}_{\min}(y),$$

which establishes statement (ii) of the theorem. $\qquad\square$

In Theorem 4.69, condition (4.90) is weaker than (4.88) and hence more update rules B will be likely to fulfill it. However, then also the resulting conclusion (4.91) which can be made is weaker than (4.89). Namely, in case (i), we can ensure that the best achievable steady-state cost along the closed-loop system converges, i.e., $\lim_{t\to\infty} \ell_{\min}(x(t))$ exists. Furthermore, κ_∞ is equal to this limit, which is at least as good as the minimum of the best robustly achievable steady-state cost on the ω-limit set of the closed-loop system. On the other hand, in case (ii), we cannot necessarily ensure that $\lim_{t\to\infty} \ell_{\min}(x(t))$ exists, but only that κ_∞ is equal to $\limsup_{t\to\infty} \ell_{\min}(x(t))$, which in turn is at least as good as the supremum of the best robustly achievable steady-state cost on the ω-limit set of the closed-loop system. Furthermore, we remark that if $\omega_B(x_0)$ is just a singleton, or, more general, if $\bar{\ell}_{\min}(y_1) = \bar{\ell}_{\min}(y_2)$ for all $y_1, y_2 \in \omega_B(x_0)$, then the right hand sides of (4.91) and (4.89), i.e., the upper bounds for κ_∞, are the same.

Remark 4.71. *For Theorem 4.69, we note the following concerning the possibility to relax Assumption 2.6 to Assumption 2.2, i.e., to allow for an unbounded constraint set $\mathbb{Z}$. One can show that Proposition 2 of (Fagiano and Teel, 2013) is still satisfied under Assumption 2.2 if one in addition assumes that ℓ is bounded on $\mathbb{Z}$, or alternatively that f and ℓ are Lipschitz continuous (or, more general, Hölder continuous) on $\mathbb{Z}$. Hence the first equalities in (4.89) and (4.91) still hold in this case. On the other hand, the second inequalities in (4.89) and (4.91) are only well defined and valid if $\mathbb{Z}$ is compact.*

Specific update rules B and their properties

In the following, we propose and discuss several different update rules B which ensure that the conditions of Theorems 4.65 and 4.69 are satisfied. Some of these update rules are actually more general than the formula (4.80), as they also include some memory of the past evolution of β. We start with rather simple update rules which lead to monotonically increasing β. To this end, let

$$\delta(t) := \kappa(t+1) - \ell_{\min}(x(t)) \overset{(4.79)}{=} \ell(x^0(N|t), u^0(N|t)) - \ell_{\min}(x(t)). \qquad (4.92)$$

- Update rule 1: $B_1(\beta(t), x(t), \kappa(t+1)) := \beta(t) + d$ for some $d > 0$.

- Update rule 2: $B_2(\beta(t), x(t), \kappa(t+1)) := \beta(t) + \alpha(\delta(t))$ for some $\alpha \in \mathcal{K}$.

The appeal of update rules 1 and 2 is clearly their simplicity. A further advantage of update rule 1 is the fact that in contrast to update rule 2, $\ell_{\min}(x(t))$ does not have to be known at each time t. However, with update rule 1, in any case $\beta \to \infty$, which might not be desirable.

Lemma 4.72. *The update rules 1 and 2 are such that (4.88) holds; furthermore, for update rule 2, Assumption 4.12 is satisfied.*

Proof. With update rule 1, (4.88) is trivially satisfied as $\liminf_{t\to\infty} \beta(t) = \infty$ independent of the behavior of $\boldsymbol{x}$ and $\boldsymbol{\kappa}$. For update rule 2, consider the following. If we have $\kappa_\infty - \liminf_{t\to\infty} \ell_{\min}(x(t)) > 0$ for some sequences $\boldsymbol{x}$ and $\boldsymbol{\kappa}$ with $\boldsymbol{\kappa}$ nonincreasing and $\lim_{t\to\infty} \kappa(t) = \kappa_\infty$, then $\delta(t_i) \geq c > 0$ for an infinite sequence of time instants $\{t_i\} \subseteq \mathbb{I}_{\geq 0}$ and some constant $c > 0$. But this immediately implies that for the corresponding sequence $\boldsymbol{\beta}$, we have $\liminf_{t\to\infty} \beta(t) = \infty$ as $\alpha \in \mathcal{K}$. Hence with both update rules 1 and 2, (4.88) is satisfied. In order to establish the second claim of the lemma, consider the following. For update rule 2, we have $\gamma(t) = \beta(t+1) - \beta(t) = \alpha(\delta(t))$. As $\boldsymbol{\kappa}$ is nonincreasing and $\ell_{\min}(x)$ is bounded from below (by $\ell(x^*, u^*) = 0$), from the definition of δ in (4.92) we obtain $\gamma(t) \leq \alpha(\kappa_0) < \infty$ for all $t \in \mathbb{I}_{\geq 0}$. Furthermore, $\beta(t) \geq \beta_0 \geq 0$ for all $t \geq 0$. Finally, as (4.88) is satisfied, by Theorem 4.69(i) we conclude that $\lim_{t\to\infty} \delta(t) = 0$ for the closed-loop system resulting from application of Algorithm 4.63, and hence also $\lim_{t\to\infty} \gamma(t) = 0$ as $\alpha \in \mathcal{K}$, which means that Assumption 4.12 is satisfied. $\qquad \square$

We now turn our attention to slightly more complex update rules which are nonmonotonic, but allow for a reset of the terminal weight β. Such nonmonotonic update rules are desirable as β might stay smaller, which, as mentioned above, might be good for performance reasons. Moreover, one might also have a greater robustness for β to stay bounded in case of disturbances (see, e.g., Example 4.79 in Section 4.3.5).

- Update rule 3: Let $\alpha_1, \alpha_2 \in \mathcal{K}$.

$$B_3(\beta(t), x(t), \kappa(t+1)) := \begin{cases} 1 & \text{if } C_3(t) \leq 0, \\ \beta(t) + \alpha_2(\delta(t)) & \text{else,} \end{cases}$$

where $C_3(0) = 0$ and for each $t \in \mathbb{I}_{\geq 1}$, $C_3(t) := \kappa(t+1) - \kappa(t_{\text{last}}) + \alpha_1(\delta(t))$ with $t_{\text{last}} := \max_{s \in \mathbb{I}_{[0,t]}, \beta(s)=1} s$.

Lemma 4.73. *Update rule 3 is such that* (4.90) *holds; furthermore, at least one of Assumptions 4.12 and 4.13 is satisfied.*

Proof. We start by proving the first claim of the lemma. If $\kappa_\infty - \limsup_{t \to \infty} \ell_{\min}(x(t)) > 0$ for some sequences x and κ with κ nonincreasing and $\lim_{t \to \infty} \kappa(t) = \kappa_\infty$, then there exists $\hat{t} > 0$ such that $\delta(t) \geq c > 0$ for all $t \geq \hat{t}$ and some $c > 0$. We will show that in this case the number of resets to 1 of the corresponding sequence β is finite. Namely, assume it was not. Then, both $\kappa(t+1)$ as well as $\kappa(t_{\text{last}})$ converge to κ_∞. But this implies that there exists $\bar{t} \geq \hat{t}$ such that $C_3(t) > 0$ for all $t \geq \bar{t}$ (as $\delta(t) \geq c > 0$), i.e., the reset condition is not satisfied anymore, which gives a contradiction. Hence the number of resets of β is finite. But then, according to the definition of B_3, we immediately obtain that $\liminf_{t \to \infty} \beta(t) = \limsup_{t \to \infty} \beta(t) = \infty$, i.e., (4.90) is satisfied. To prove the second claim of the lemma, consider the following. First, the definition of B_3 implies that $\beta(t) \geq \min\{\beta_0, 1\} \geq 0$ for all $t \in \mathbb{I}_{\geq 0}$. Second, whenever $C_3(t) \leq 0$, we have $\gamma(t) \leq 1$ (as $\beta(t) \geq 0$ for all $t \in \mathbb{I}_{\geq 0}$); if $C_3(t) > 1$, we again obtain $\gamma(t) = \alpha_2(\delta(t)) \leq \alpha_2(\kappa_0)$ as established in the proof of Lemma 4.72. Hence $\gamma(t) \leq \max\{1, \alpha_2(\kappa_0)\} < \infty$ for all $t \in \mathbb{I}_{\geq 0}$. Thus, in order to show that at least one of Assumptions 4.12 and 4.13 is satisfied, it remains to show that either $\limsup_{t \to \infty} \gamma(t) \leq 0$ or $\liminf_{t \to \infty} \beta(t) < \infty$ for the closed-loop sequence β resulting from application of Algorithm 4.63. The latter is immediately satisfied if the number of resets of β is infinite, i.e., $C_3(t_i) \leq 0$ for an infinite sequence $\{t_i\} \subseteq \mathbb{I}_{\geq 0}$. Now assume for contradiction that both $\limsup_{t \to \infty} \gamma(t) > 0$ and the number of resets of β is finite, i.e., there exists $\check{t} \in \mathbb{I}_{\geq 0}$ such that $C_3(t) > 0$ for all $t \in \mathbb{I}_{\geq \check{t}}$. Then, the definition of B_3 immediately implies that β is monotonically increasing for $t \geq \check{t}$ with $\liminf_{t \to \infty} \beta(t) = \infty$. As shown in the proof of Theorem 4.69(i), this would result in (4.89). But the first equality in (4.89) implies that $\lim_{t \to \infty} \delta(t) = 0$ and hence $\limsup_{t \to \infty} \gamma(t) \leq 0$, which contradicts our assumption. Hence at least one of Assumptions 4.12 and 4.13 is satisfied, which concludes the proof of Lemma 4.73. □

Remark 4.74. *With update rule 3, the number of resets of the closed-loop sequence β can very well be infinite. In the proof of Lemma 4.73, we only showed that if $\kappa_\infty - \limsup_{t \to \infty} \ell_{\min}(x(t)) > 0$ for some sequences x and κ, which are regarded as* open-loop *inputs in (4.80), then the number of resets of the corresponding sequence β is finite. However, this situation does not occur for the* closed-loop *sequences x and κ resulting from application of Algorithm 4.63, but $\kappa_\infty - \limsup_{t \to \infty} \ell_{\min}(x(t)) = 0$ according to Theorem 4.69(ii).*

In both update rules 2 and 3, one has to know $\ell_{\min}(x(t))$ at each time t. This might be a rather strict requirement as one has to solve an additional optimization problem at each time step in order to determine $\ell_{\min}(x(t))$. Hence in the following, we consider update rules where $\ell_{\min}(x(t))$ does not have to be calculated at each time t, but only on an infinite subsequence $\{t_i\} \subseteq \mathbb{I}_{\geq 0}$. Hence these update rules are implementable if one can ensure that one persistently has the computational power to solve this additional optimization problem at *some* time instant.

Let $\nu : \mathbb{I}_{\geq 0} \to \{0, 1\}$ be any sequence with $\nu(0) = 1$ and $\limsup_{t \to \infty} \nu(t) = 1$, and for each $t \in \mathbb{I}_{\geq 0}$, let $\tau_\nu(t) := \max_{s \in \mathbb{I}_{[0,t]}, \nu(s)=1} s$. Furthermore, let $\alpha_1, \alpha_2 \in \mathcal{K}$ and define

$$\bar{\delta}(t) := \max\{\kappa(t+1) - \ell_{\min}(x(\tau_\nu(t))), 0\}. \tag{4.93}$$

- Update rule 4: $B_4(\beta(t), x(t), \kappa(t+1)) := \beta(t) + \alpha_1(\bar{\delta}(t))$.

- Update rule 5:

$$B_5(\beta(t), x(t), \kappa(t+1)) := \begin{cases} 1 & \text{if } C_5(t) \leq 0, \\ \beta(t) + \alpha_2(\bar{\delta}(t)) & \text{else,} \end{cases}$$

where $C_5(0) = 0$ and for each $t \in \mathbb{I}_{\geq 1}$, $C_5(t) := \kappa(t+1) - \kappa(t_{\text{last}}) + \alpha_1(\bar{\delta}(t))$ with $t_{\text{last}} := \max_{s \in \mathbb{I}_{[0,t]}, \beta(s)=1} s$.

Lemma 4.75. *Update rules 4 and 5 are such that* (4.90) *holds. Moreover, for update rule 4, Assumption 4.12 is satisfied, whereas at least one of Assumptions 4.12 and 4.13 is satisfied for update rule 5.*

Proof. The proof is similar to the proof of Lemma 4.73. We start proving the first claim. If $\kappa_\infty - \limsup_{t \to \infty} \ell_{\min}(x(t)) > 0$ for some sequences $\boldsymbol{x}$ and $\boldsymbol{\kappa}$ with $\boldsymbol{\kappa}$ nonincreasing and $\lim_{t \to \infty} \kappa(t) = \kappa_\infty$, then, as $\limsup_{t \to \infty} \nu(t) = 1$, there exists $\hat{t} \in \mathbb{I}_{\geq 0}$ such that $\bar{\delta}(t) \geq c > 0$ for all $t \in \mathbb{I}_{\geq \hat{t}}$ and some $c > 0$. For update rule 4, this immediately implies that $\liminf_{t \to \infty} \beta(t) = \limsup_{t \to \infty} \beta(t) = \infty$. For update rule 5, one can show with the same reasoning as in the proof of Lemma 4.73 that the number of resets of $\boldsymbol{\beta}$ in this case is finite, which again yields $\liminf_{t \to \infty} \beta(t) = \limsup_{t \to \infty} \beta(t) = \infty$. Hence with both update rules 4 and 5, (4.90) is satisfied. The second statement of the lemma can be proven analogous to Lemma 4.73. Namely, for the closed-loop sequence $\boldsymbol{\beta}$ resulting from application of Algorithm 4.63, one can show by contradiction that $\limsup_{t \to \infty} \gamma(t) \leq 0$ for update rule 4, and for update rule 5 that either $\limsup_{t \to \infty} \gamma(t) \leq 0$ or the number of resets is infinite, which implies that $\liminf_{t \to \infty} \beta(t) = 1 < \infty$. Finally, the upper bound for $\gamma(t)$ can be calculated as in the proof of Lemmas 4.72 and 4.73, respectively, and again by definition $\beta(t) \geq 0$ for all $t \in \mathbb{I}_{\geq 0}$. Hence for update rule 4, Assumption 4.12 is satisfied, whereas at least one of Assumptions 4.12 and 4.13 is satisfied for update rule 5, as claimed. $\square$

Next, we would like to find an update rule which is non-monotonic but still satisfies the stronger condition of statement (i) of Theorem 4.69, i.e. (4.88). To this end, consider the following update rule, which is a slight modification of update rule 3.

- Update rule 6: Let $\alpha_1, \alpha_2 \in \mathcal{K}$.

$$B_6(\beta(t), x(t), \kappa(t+1)) := \begin{cases} 1 & \text{if } C_6(t) \leq 0, \\ \beta(t) + \alpha_2(\delta(t)) & \text{else,} \end{cases}$$

where $C_6(0) = 0$ and for each $t \in \mathbb{I}_{\geq 1}$, $C_6(t) := \kappa(t+1) - \kappa(t_{\text{last}}) + \alpha_1(\max_{s \in \mathbb{I}_{[t_{\text{last}},t]}} \delta(s))$ with $t_{\text{last}} := \max_{s \in \mathbb{I}_{[0,t]}, \beta(s)=1} s$.

Lemma 4.76. *Update rule 6 is such that (4.88) holds; moreover, Assumption 4.12 is satisfied.*

Proof. We again start by proving the first claim. If $\kappa_\infty - \liminf_{t \to \infty} \ell_{\min}(x(t)) > 0$ for some sequences $\boldsymbol{x}$ and $\boldsymbol{\kappa}$ with $\boldsymbol{\kappa}$ nonincreasing and $\lim_{t \to \infty} \kappa(t) = \kappa_\infty$, then $\delta(t_i) \geq c > 0$ for an infinite sequence of time instants $\{t_i\} \subseteq \mathbb{I}_{\geq 0}$ and some constant $c > 0$. Again, we show that in this case the number of resets of the corresponding sequence $\boldsymbol{\beta}$ is finite. Namely, assume it was not. Then, both $\kappa(t+1)$ as well as $\kappa(t_{\text{last}})$ converge to κ_∞, i.e., there exists $\hat{t} > 0$ such that $\kappa(t+1) - \kappa(t_{\text{last}}) \geq -\alpha_1(c/2)$ for all $t \geq \hat{t}$. Now consider the next time instant $t'_i \geq \hat{t}$ such that $\delta(t'_i) \geq c$. Then, for all $t \geq t'_i$, the reset condition for $\boldsymbol{\beta}$ will not be satisfied anymore as (by induction) we obtain

$$C_6(t) \geq -\alpha_1(c/2) + \alpha_1(\max_{s \in \mathbb{I}_{[t_{\text{last}},t]}} \delta(s)) \geq -\alpha_1(c/2) + \alpha_1(c) > 0,$$

where the last inequality follows as $\alpha_1 \in \mathcal{K}$. Hence the number of resets of $\boldsymbol{\beta}$ is finite. But then the definition of B_6 yields that $\liminf_{t \to \infty} \beta(t) = \infty$, i.e., (4.88) is satisfied. To show that Assumption 4.12 is satisfied, the upper bound for $\gamma(t)$ can be calculated as in the proof of Lemma 4.73, and again by definition $\beta(t) \geq 0$ for all $t \in \mathbb{I}_{\geq 0}$; furthermore, as (4.88) holds, by Theorem 4.69(i) we conclude that $\lim_{t \to \infty} \delta(t) = 0$ for the closed-loop system resulting from application of Algorithm 4.63, and hence $\limsup_{t \to \infty} \gamma(t) \leq 0$ due to the definition of B_6. $\qquad\square$

Remark 4.77. *In update rules 2-6, the $\mathcal{K}$-functions α (respectively, α_1 and α_2) are tuning parameters determining how fast the terminal weight sequence $\boldsymbol{\beta}$ grows and how often it is reset. If these $\mathcal{K}$-functions are "flat enough", $\boldsymbol{\beta}$ will grow more slowly and will be reset more often, respectively. We will further illustrate this property with Example 4.79 in Section 4.3.5.*

Remark 4.78. *In update rules 3, 5 and 6, one could also reset $\boldsymbol{\beta}$ to some value $\hat{\beta} \geq 0$ different than 1. Resetting $\boldsymbol{\beta}$ to 1 is a canonical choice, as with $\beta(t) = 1$ the terminal state/input pair is weighted in the same way as all other state/input pairs in the cost function (4.78).*

4.3.4 Discussion

In this section, we comment and discuss various properties of the results obtained in Sections 4.3.1–4.3.3. First, Theorem 4.65 established that κ_∞ is an upper bound for the closed-loop average performance for possibly time-varying terminal weights $\beta(t)$. In general, κ_∞ depends both on the initial values κ_0 and x_0 as well as on the update rule B for the terminal weight. As the sequence $\boldsymbol{\kappa}$ is nonincreasing and bounded from below by $\ell(x^*, u^*)$, immediate a priori bounds for κ_∞ are $\ell(x^*, u^*) \leq \kappa_\infty \leq \kappa_0$. Hence it would be tempting to choose κ_0 as small as possible. However, a small value of κ_0 might result in a degradation of the (transient) performance, and moreover, the feasible region critically depends on κ_0. Namely, if we choose $\kappa_0 = \ell(x^*, u^*)$, then (4.77d) reduces to $x(N|t) \in S^*$ for all $t \in \mathbb{I}_{\geq 0}$, which in case of a unique optimal state/input equilibrium pair (x^*, u^*), i.e.,

$S^* = \{(x^*, u^*)\}$, implies that the feasible region $\mathbb{X}_N(0)$ is that of a fixed terminal point setting. Furthermore, $\mathbb{X}_N(0)$ is empty if $\kappa_0 < \ell(x^*, u^*)$. On the other hand, the largest feasible region is obtained if κ_0 is chosen such that $\kappa_0 \geq \max_{(x,u) \in \mathbb{Z}, x=f(x,u)} \ell(x, u)$ in case that $\mathbb{Z}$ is compact (or $\kappa_0 = \infty$ otherwise), as then each steady-state in $\mathbb{Z}$ can serve as a terminal state at time 0. Hence a good tradeoff for κ_0 has to be found.

Theorem 4.69 shows that κ_∞ can be upper bounded in terms of the best robustly achievable steady-state cost of points in the ω-limit set of the closed-loop system (which in general depends on the specific update rule used). Each of the presented update rules has different advantages concerning simplicity of implementation and strength of the results which can be obtained. According to Theorems 4.65 and 4.69, the strongest results can be proven for update rules 2 and 6, as both Assumption 4.12 and (4.88) are satisfied. On the other hand, the weakest results[18] are obtained for update rules 3 and 5, as only (4.90) and possibly only Assumption 4.13 are satisfied. Regarding implementation issues, update rules 2 and 4 would be beneficial for their simplicity, as no reset condition has to be checked. On the other hand, as already discussed, resetting β is beneficial in order not to let the terminal weight grow unnecessarily big, which might degrade performance. Furthermore, as already mentioned above, a big advantage of update rules 4 and 5 is that $\ell_{\min}(x(t))$ does not have to be calculated at each time step t, but only on an infinite subsequence $\{t_i\} \subseteq \mathbb{I}_{\geq 0}$. Note that in the special case where $\ell_{\min}(x) = \ell(x^*, u^*)$ for all $x \in \mathbb{X}_N(0)$ (such as in Example 4.79 in Section 4.3.5), update rules 2 and 4 as well as 3 and 5 coincide, as $\delta(t) = \bar{\delta}(t)$ for all $t \in \mathbb{I}_{\geq 0}$ in this case, and $\ell_{\min}(x)$ does not have to be calculated at all online. Finally, we remark that while update rule 6 yields, as just discussed, the strongest results of those update rules exhibiting resets, its reset condition is also the strictest one, and hence in general resets do not occur as frequently compared to update rules 3 and 5. In summary, when choosing a specific update rule B for the terminal weight, several different aspects have to be traded off against each other such as to decide which one is best suited for the specific problem considered.

Theorems 4.65 and 4.69 guarantee that the closed-loop asymptotic average performance is upper bounded by the best robustly achievable steady-state cost of points in the ω-limit set of the closed-loop state sequence. While this is a desirable behavior, note that this bound is rather of conceptual nature, in the sense that it can only be verified a posteriori. In our recent work (Müller et al., 2014a,e), we showed that improved and a priori verifiable bounds on the closed-loop asymptotic average performance can be obtained if a generalized terminal region constraint instead of the generalized terminal equality constraint (4.77d) is used. Namely, if the generalized terminal region is appropriately defined, one can show that κ_∞ is equal to a local minimum of the stage cost function ℓ restricted to the steady-state set S. In case of linear systems with convex cost and constraints, this means that $\kappa_\infty = \ell(x^*, u^*)$, which recovers the average performance result of Theorem 2.5, i.e., for economic MPC with a fixed terminal constraint. However, we note that for general nonlinear systems, the design of such a generalized terminal region might be difficult, while the setting presented in this section, i.e., using a generalized terminal equality constraint, is readily applicable. For further details concerning the extension using a generalized terminal region, we refer the interested reader to (Müller et al., 2014a,e).

---

[18]For update rule 1, which clearly is the most simple update rule and which is rather stated for motivation of the subsequent ones, Theorem 4.65 cannot be evoked at all, as neither Assumption 4.12 nor Assumption 4.13 is satisfied.

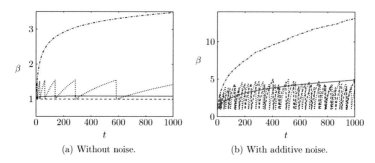

(a) Without noise. (b) With additive noise.

Figure 4.7: Closed-loop sequences β in Example 4.79 using different update rules: update rule 2 (respectively 4) with $\alpha(r) = r$ (dash-dotted); update rule 2 (respectively 4) with $\alpha(r) = r^2$ (solid); update rule 3 (respectively 5) with $\alpha_1(r) = \alpha_2(r) = r$ (dotted); update rule 3 (respectively 5) with $\alpha_1(r) = r^2$ and $\alpha_2(r) = r$ (dashed).

4.3.5 Numerical examples

Example 4.79. *Consider the scalar system $x(t+1) = (1 - u(t))x(t)$ with state and input constraint set $\mathbb{Z} = \mathbb{X} \times \mathbb{U} = [0,1]^2$, cost function $\ell(x,u) = x + (1/2)u^2$ and prediction horizon $N = 1$. The unique optimal state/input equilibrium pair is given by $(x^*, u^*) = (0,0)$, and for all $x \in \mathbb{X}$, we obtain $\ell_{\min}(x) = \ell(x^*, u^*) = 0$. It is straightforward to calculate that if $\kappa_0 \geq \ell(x_0, 0)$, the minimizer of Problem 4.62 with $N = 1$ is given by $u^0(0|t) = \min\{\beta(t)x(t), 1\}$, $u^0(1|t) = 0$, for all $t \in \mathbb{I}_{\geq 0}$. The closed-loop system resulting from application of Algorithm 4.63 is then given by $x(t+1) = (1 - \min\{\beta(t)x(t), 1\})x(t)$, with $\beta(t)$ according to the specific used update rule. For each of the above discussed update rules and each initial condition $x_0 \in \mathbb{X}$ and $\beta_0 \geq 0$, the ω-limit set is given by $\omega_B(x_0) = \{0\}$, and $\max_{y \in \omega_B(x_0)} \overline{\ell}_{\min}(y) = \min_{y \in \omega_B(x_0)} \overline{\ell}_{\min}(y) = 0$. Hence for each of the above update rules, by Theorem 4.69 we conclude that $\kappa_\infty = 0$. However, the closed-loop evolution of β is quite different when using different update rules (see Figure 4.7(a)). As noted in Section 4.3.4, update rules 2 and 4 as well as 3 and 5 coincide, as $\ell_{\min}(x) = \ell(x^*, u^*)$ for all $x \in \mathbb{X}$. For update rule 2 (respectively 4), one can show that when using $\alpha(r) = r$, β also does not stay bounded, as x converges to zero very slowly (dash-dotted curve in Figure 4.7(a)). On the other hand, when using $\alpha(r) = r^2$, one observes that β stays bounded and converges (solid curve in Figure 4.7(a)). When using update rule 3 (respectively 5) with $\alpha_1(r) = r^2$ and $\alpha_2(r) = r$, one can show that $\beta \equiv 1$, as the reset condition is always fulfilled (dashed curve in Figure 4.7(a)). For $\alpha_1(r) = \alpha_2(r) = r$, on the other hand, one observes kind of a sawtooth behavior of β (dotted curve in Figure 4.7(a)).*

As discussed earlier, an advantage of nonmonotonic update rules is that they might be more robust to disturbances. Namely, consider again the same update rules as used in Figure 4.7(a), but now an additive random disturbance (uniformly distributed over the interval $[0, 0.2]$) acts on the system. As one can see from the simulations (see Figure 4.7(b)), the monotonic update rule 2 (respectively 4) now leads to an unbounded β, for both the choices $\alpha(r) = r$ and $\alpha(r) = r^2$. On the other hand, update rule 3 (respectively 5) leads to a bounded β. □

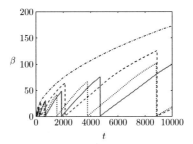

Figure 4.8: Closed-loop sequences β in Example 4.80 for different update rules: update rule 2 with $\alpha(r) = r$ (dash-dotted); update rule 3 with $\alpha_1(r) = \alpha_2(r) = r$ (solid); update rule 5 with $\alpha_1(r) = \alpha_2(r) = r$ and $\nu(t) = 1$ if $t = 50k$, $k \in \mathbb{I}_{\geq 0}$, and $\nu(t) = 0$ otherwise (dotted); update rule 6 with $\alpha_1(r) = \alpha_2(r) = r$ (dashed).

Example 4.80. *Consider the scalar system $x(t + 1) = x(t)/(1 + x(t)u(t))$ with state and input constraint set $\mathbb{Z} = \mathbb{X} \times \mathbb{U} = [0,1]^2$, cost function $\ell(x,u) = x^2 + 2u + xu^2$ and prediction horizon $N = 1$. For all $x \in \mathbb{X}$, the best state/input equilibrium pair reachable in one step is $(x/(x+1), 0)$, i.e., $\ell_{\min}(x) = x^2/(x+1)^2$ for all $x \in \mathbb{X}$. Furthermore, the unique optimal state/input equilibrium pair is given by $(x^*, u^*) = (0,0)$. It is straightforward to calculate that if $x_0 > 0$ and $\kappa_0 \geq \ell(x_0, 0)$, the minimizer of Problem 4.62 with $N = 1$ is given by*

$$u^0(0|t) = \min\left\{1, \max\left\{\frac{\sqrt[4]{\beta(t)x(t)^3} - 1}{x(t)}, 0\right\}\right\}, \qquad u^0(1|t) = 0$$

for all $t \in \mathbb{I}_{\geq 0}$. The closed-loop system resulting from application of Algorithm 4.63 is then given by

$$x(t+1) = \frac{x(t)}{1 + x(t)\min\left\{1, \max\left\{\frac{\sqrt[4]{\beta(t)x(t)^3} - 1}{x(t)}, 0\right\}\right\}}, \tag{4.94}$$

with $\beta(t)$ according to the specific used update rule. From (4.94), it follows that if β is bounded, i.e., $\beta(t) \leq \hat{\beta}$ for some $\hat{\beta} < \infty$ and all $t \in \mathbb{I}_{\geq 0}$ (which in particular would be the case if a constant terminal weight was used), then x remains constant once $x \leq 1/\sqrt[3]{\hat{\beta}}$. So in order to converge to the optimal steady-state and to obtain $\kappa_\infty = 0$, we necessarily need that $\limsup_{t\to\infty} \beta(t) = \infty$. This happens for all six update rules proposed in Section 4.3.3, as according to Lemmas 4.72–4.76, for each of these update rules either (4.88) or (4.90) is satisfied. Figure 4.8 shows closed-loop evolutions of β for different update rules.

Example 4.81. *The following example is a simple illustration of the fact why in Theorem 4.69, we only can ensure that κ_∞ converges to a value not greater than the minimum (respectively, maximum) of the best robustly achievable steady-state cost on the ω-limit set. Namely, consider the two-dimensional system*

$$x_1(t+1) = x_1(t) + u(t),$$
$$x_2(t+1) = x_2(t) - |\sin(x_2(t))| + u(t),$$

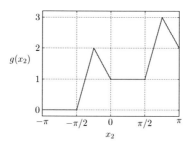

Figure 4.9: The function $g(x_2)$ in Example 4.81.

with $x = [x_1 \ x_2]^T \in \mathbb{R}^2$ and $u \in \mathbb{R}$. The state and input constraint set is given by $\mathbb{Z} = \mathbb{X} \times \mathbb{U}$, where $\mathbb{X} = \mathbb{R} \times [-\pi, \pi]$ and $\mathbb{U} = [-u_{max}, u_{max}]$ with u_{max} such that $2u_{max} + \sin(u_{max}) = \pi$. The set of all feasible state/input equilibrium pairs is given by $S = \{(x, u) : x_1 \in \mathbb{R}, x_2 \in \{\pm\pi, 0\}, u = 0\}$. Furthermore, the prediction horizon is $N = 2$, and the cost function is $\ell(x, u) = |u| + g(x_2)$ with $g(x_2)$ as depicted in Figure 4.9. Note that the parameter u_{max} appearing in the input constraint set $\mathbb{U}$ is defined such that for each $x \in \mathbb{X}$ with $x_2 \in (0, \pi]$, the best reachable steady-state in two steps is such that $x_2 = 0$, and for each $x \in \mathbb{X}$ with $x_2 \in [-\pi, 0]$, the best reachable steady-state in two steps is such that $x_2 = -\pi$. Thus, for all $x_1 \in \mathbb{R}$ and each $\varepsilon > 0$ small enough, $\ell'_{\min}([x_1 \ 0]^T; \varepsilon) = \ell([x_1 \ 0]^T, 0) = g(0) = 1$, and hence also $\bar{\ell}_{\min}([x_1 \ 0]^T) = 1$. On the other hand, $\ell_{\min}([x_1 \ 0]^T) = \ell([x_1 \ -\pi]^T, 0) = g(-\pi) = 0 < \bar{\ell}_{\min}([x_1 \ 0]^T)$, which means that $\ell'_{\min}([x_1 \ 0]^T; \varepsilon)$ is not continuous in ε at $\varepsilon = 0$.

Straightforward calculations yield that if $x_0 \in \mathbb{R} \times (0, \pi/2]$ and $\kappa_0 \geq \ell([x_1 \ 0]^T, 0)$, the minimizer of Problem 4.62 is, independent of $\beta(t)$, given by

$$u^0(0|t) = u^0(2|t) = 0$$
$$u^0(1|t) = -x_2(t) + |\sin(x_2(t))| + |\sin(x_2(t) - |\sin(x_2(t))|)|$$

for all $t \in \mathbb{I}_{\geq 0}$. This means that the closed-loop system resulting from application of Algorithm 4.63 satisfies $\lim_{t\to\infty} x_2(t) = 0$ and $x_1(t) = x_1(0)$ for all $t \in \mathbb{I}_{\geq 0}$, and hence $\omega_B(x_0) = \{[x_1(0) \ 0]^T\}$ independent of the closed-loop behavior of β. Furthermore, $x_2^0(N|t) = 0$ for all $t \in \mathbb{I}_{\geq 0}$, and hence $\kappa_\infty = \ell([x_1(0) \ 0]^T, 0) = \bar{\ell}_{\min}([x_1(0) \ 0]^T) > \ell_{\min}([x_1(0) \ 0]^T)$. In fact, since $\lim_{t\to\infty} x_2(t) = 0$ and $u(t) = u^0(0|t) = 0$ for all $t \in \mathbb{I}_{\geq 0}$, it follows that (4.81) holds with equality, i.e., the average performance of the closed-loop system is equal to the best robustly achievable steady-state cost of the ω-limit set, $\bar{\ell}_{\min}([x_1(0) \ 0]^T)$, which is strictly greater than the best achievable steady-state cost of the ω-limit set, $\ell_{\min}([x_1(0) \ 0]^T)$.

4.4 Summary

In this chapter, several contributions in the field of economic MPC were obtained. We firstly provided an in-depth analysis of a certain dissipativity property which has turned out to play a crucial role in economic MPC. In particular, we established its necessity for optimal steady-state operation and robustness properties with respect to changes in the constraints. Secondly, we developed economic MPC schemes which guarantee satisfaction of both asymptotic and transient average constraints, which are of interest in economic MPC as the closed-loop system is not necessarily convergent. We also provided a Lyapunov-like convergence analysis for economic MPC with average constraints, given that a certain relaxed dissipativity condition is satisfied. Thirdly, we developed an economic MPC scheme with self-tuning terminal cost. We formulated conditions which the self-tuning update rule has to satisfy in order to guarantee certain closed-loop performance bounds, and we proposed and discussed several specific update rules satisfying these conditions.

The results presented in this chapter can serve as a starting point for various interesting future research topics. For a detailed discussion on these issues, we refer the reader to Section 6.2 at the end of this thesis.

Chapter 5

Distributed economic MPC

In this chapter, we combine and unify the ideas developed in the previous two chapters and present a distributed economic MPC scheme for cooperative control of self-interested systems. In Chapter 3, we translated general cooperative control tasks into the stabilization of some set $\mathbb{X}^*$, and the sum of the individual cost functions was assumed to be positive definite with respect to that set, i.e., had to be consistent with the cooperative control task to be achieved. This means that the resulting distributed MPC schemes were formulated in the setting of stabilizing MPC. In contrast, in this chapter we develop a distributed *economic* MPC scheme. Namely, each system uses its own, individual, general (economic) objective function, which models its self-interest and which may be conflicting with the overall cooperative control task to be achieved. This cooperative control objective is then formulated in terms coupling constraints. In such a setting, economic MPC is a natural candidate and well suited for tackling these kind of cooperative control problems. We note that a similar setup was also considered by Driessen et al. (2012), where each system uses a modified economic cost function in order to achieve stabilization of the coupling output, and the overall optimal steady-state has to be known a priori, which is then used as a terminal constraint. The latter is also the case in the work of (Lee and Angeli, 2011, 2012; Wang, 2011), which, as mentioned in Section 1.2.1, extend the work of Stewart et al. (2010) to an economic setting.

In contrast, our basic assumption is that the overall optimal steady-state (including the coupling constraints) is *not* known a priori, but has to be negotiated between the systems online by implementing some distributed optimization algorithm. This assumption is a crucial requirement in large scale networks, and furthermore in situations where the number of systems in the network changes online (see Section 5.4 for a more detailed discussion of the latter issue). We assume that communication between the systems requires time, i.e., the systems already have to perform control actions while still negotiating with their neighboring systems. This premise was also adopted in a similar context by Zelazo et al. (2013), where a shrinking-horizon preference agreement algorithm was developed for scalar single integrator systems with quadratic objectives.

The remainder of this chapter is structured as follows. In Section 5.1, we state the detailed problem setup and present the distributed economic MPC scheme. As indicated above, the proposed control structure will be hierarchical in the sense that it consists of a distributed optimization algorithm to determine the overall optimal steady-state, whose current iterates at each time step are then used to determine a suitable terminal constraint for the economic MPC problem solved by each system. In Section 5.2, a convergence analysis of the proposed distributed economic MPC scheme is given, and Section 5.3 illustrates the obtained results by considering the problem of synchronizing several agents with conflicting objective. Finally, Section 5.4 gives a brief summary and discussion of the obtained results.

The results presented in this chapter are based on (Müller and Allgöwer, 2014b); furthermore, we note that Lemma 5.10 is a slight extension of Lemma 1 in (Müller et al., 2014a,e).

5.1 Problem setup and distributed economic MPC scheme

Consider again the setting of Chapter 3, i.e., a network of P dynamically decoupled discrete-time systems. In this chapter we assume that the system dynamics (3.1) are linear, i.e., given by

$$x_i(t+1) = A_i x_i(t) + B_i u_i(t), \qquad x_i(0) = x_{i0} \tag{5.1}$$

with $t \in \mathbb{I}_{\geq 0}$ and $i \in \mathbb{I}_{[1,P]}$, and the pair (A_i, B_i) is assumed to be stabilizable. Each system is again subject to the pointwise-in-time local state and input constraints (3.2), i.e., $(x_i(t), u_i(t)) \in \mathbb{Z}_i \subseteq \mathbb{X}_i \times \mathbb{U}_i$ for all $t \in \mathbb{I}_{\geq 0}$. The set of all feasible state/input equilibrium pairs for system i is defined as $\mathcal{S}_i := \{(x_i, u_i) \in \mathbb{Z}_i : x_i = A_i x_i + B_i u_i\}$. Denote again by $x(t)$ and $u(t)$ the state and input of the overall system at time t, i.e., $x(t) := [x_1(t)^T \ \ldots \ x_P(t)^T]^T \in \mathbb{R}^{n_{\mathrm{ov}}}$ (with $n_{\mathrm{ov}} = \sum_{i=1}^{P} n_i$) and $u(t) := [u_1(t)^T \ \ldots \ u_P(t)^T]^T \in \mathbb{R}^{m_{\mathrm{ov}}}$ (with $m_{\mathrm{ov}} = \sum_{i=1}^{P} m_i$), and let $\mathcal{S} := \mathcal{S}_1 \times \cdots \times \mathcal{S}_P$. Each system is equipped with an objective function $\ell_i : \mathbb{Z}_i \to \mathbb{R}$, which can be a general (economic) cost and which models the self-interest of each agent. In this chapter, the cooperative requirement which the systems have to fulfill is translated into asymptotic satisfaction of some coupling constraints of the form

$$x \in \mathcal{C} \tag{5.2}$$

for some set $\mathcal{C} \subseteq \mathbb{R}^{n_{\mathrm{ov}}}$, i.e., $\lim_{t \to \infty} |x(t)|_{\mathcal{C}} = 0$. Note that as discussed in Chapter 1, various cooperative control tasks fit into this framework, such as the synchronization of several agents with conflicting objective (see also Section 5.3), or output agreement problems in general.

Remark 5.1. *For clarity of presentation, in this chapter we consider coupling constraints (5.2) involving only the system states x_i; nevertheless, in a similar way, also coupling constraints involving both the system states x_i and the system inputs u_i can be treated.* □

We impose the following assumption on the constraint sets $\mathbb{Z}_i$ and $\mathcal{C}$, as well as the cost functions ℓ_i.

Assumption 5.1. *For all $i \in \mathbb{I}_{[1,P]}$, the sets $\mathbb{Z}_i$ are compact and convex and ℓ_i is continuous and convex. Furthermore, the coupling constraint set $\mathcal{C}$ is closed and convex.*

Analogous to Section 2.2, define the set of overall optimal state/input equilibrium pairs (including coupling constraints) as

$$S^* := \{(y, w) \in S : y \in \mathcal{C} \text{ and } \ell(y, w) = \min_{(x,u) \in \mathcal{S}, x \in \mathcal{C}} \sum_{i=1}^{P} \ell_i(x_i, u_i)\} \tag{5.3}$$

In the following, denote again by (x^*, u^*) an arbitrary element of S^*, i.e., one of the (possibly multiple) overall optimal state/input equilibrium pairs[1], where $x^* = [(x_1^*)^T \ \ldots \ (x_P^*)^T]^T$ and $u^* = [(u_1^*)^T \ \ldots \ (u_P^*)^T]^T$. Our basic prerequisite is that (x^*, u^*) is not known a priori, but has to be calculated online via a distributed optimization algorithm, which requires communication between the systems. We assume that this communication requires time, i.e., an update step in the distributed optimization algorithm is not instantaneous; in particular, we assume that each iteration of the distributed optimization algorithm solving the minimization problem in (5.3) corresponds to one actual time step t in the evolution of the systems (5.1). This means that the systems "negotiate" about the overall optimal steady-state while already taking control actions. Let $\zeta(t) := (\xi(t), \eta(t))$ denote the iterate at time t of the distributed optimization algorithm solving the minimization problem in (5.3), where $\xi(t) := [\xi_1(t)^T \ \ldots \ \xi_P(t)^T]^T$ and $\eta(t) := [\eta_1(t)^T \ \ldots \ \eta_P(t)^T]^T$ are the state and input components of $\zeta(t)$, respectively. We then impose the following assumption.

Assumption 5.2. *The distributed optimization algorithm solving the minimization problem in (5.3) is such that* $\lim_{t \to \infty} \zeta(t) = (x^*, u^*)$. □

Remark 5.2. *Thanks to convexity of problem (5.3), many distributed optimization algorithms exist satisfying Assumption 5.2. In settings where the coupling constraint set $\mathcal{C}$ consists of several coupling constraints each of which involves only a limited number of systems, dual subgradient methods (Ruszczyński, 2006) are, e.g., well suited; for problems where coupling constraints involve all systems, one can, for example, use the recently proposed cutting-plane consensus algorithm (Bürger et al., 2014).*

Remark 5.3. *When defining the set of overall optimal state/input equilibrium pairs S^* in (5.3), one could also use different weighting factors $a_i > 0$ for each cost function ℓ_i, which would correspond to a certain prioritization of the systems.*

As discussed above, the systems already have to take control actions while still negotiating about the overall optimal steady-state. The local control actions for each system are determined in an economic MPC fashion, i.e., at each time $t \in \mathbb{I}_{\geq 0}$ with measured state $x_i(t)$, system i solves the following optimization problem, whose components will be explained in more detail below:

Problem 5.4.

$$\underset{\boldsymbol{u}_i(t)}{minimize} \ J_{i,N}(x_i(t), \boldsymbol{u}_i(t)) \tag{5.4}$$

subject to

$$x_i(0|t) = x_i(t), \tag{5.5a}$$

$$x_i(k+1|t) = A_i x_i(k|t) + B_i u_i(k|t), \quad k \in \mathbb{I}_{[0,N-1]}, \tag{5.5b}$$

$$(x_i(k|t), u_i(k|t)) \in \mathbb{Z}_i, \quad k \in \mathbb{I}_{[0,N-1]}, \tag{5.5c}$$

$$x_i(N|t) \in \mathbb{X}_i^f(x_i^s(t), t), \tag{5.5d}$$

$$\sum_{k=0}^{N-1} h_i(x_i(k|t), u_i(k|t), t) \in \mathbb{Y}_{i,t}, \tag{5.5e}$$

---

[1]In case that ℓ is assumed to be strictly convex instead of only convex, (x^*, u^*) is guaranteed to be unique.

where

$$J_{i,N}(x_i(t), \boldsymbol{u}_i(t)) := \sum_{k=0}^{N-1} \ell_i(x_i(k|t), u_i(k|t)) + V_i^f(x_i(N|t), x_i^s(t)). \tag{5.6}$$

In this problem, $\boldsymbol{u}_i(t) := \{u_i(0|t), \ldots, u_i(N-1|t)\}$ and $\boldsymbol{x}_i(t) := \{x_i(0|t), \ldots, x_i(N|t)\}$ are again input and corresponding state sequences predicted at time t over the prediction horizon $N \in \mathbb{I}_{\geq 0}$. Furthermore, as before, the minimizer of Problem 5.4 is denoted by $\boldsymbol{u}_i^0(t) := \{u_i^0(0|t), \ldots, u_i^0(N-1|t)\}$, the corresponding state sequence by $\boldsymbol{x}_i^0(t) := \{x_i^0(0|t), \ldots, x_i^0(N|t)\}$, and the corresponding optimal value function by $J_{i,N}^0(x_i(t))$. Note that for each $i \in \mathbb{I}_{[1,P]}$, Problem 5.4 is an economic MPC problem including average constraints as presented in Section 4.2.1, i.e., corresponds to Problem 4.23, except for the following difference. While in Problem 4.23, the (time-varying) terminal region $\mathbb{X}^f(t)$ was built around the optimal steady-state x^* for all $t \in \mathbb{I}_{\geq 0}$, we now use a terminal region $\mathbb{X}_i^f(x_i^s(t), t)$ which is built around the (possibly time-varying) steady-state $x_i^s(t)$, which is used to account for the fact that the overall optimal steady-state (x^*, u^*) is unknown a priori, but is calculated online as noted above. We will discuss in the following how $x_i^s(t)$ can be appropriately defined. Furthermore, the average constraint (5.5e) will be used later on to ensure convergence of the overall closed-loop system to x^*, and we show below how the functions h_i and the sets $\mathbb{Y}_{i,t}$ can be appropriately defined to this end.

Remark 5.5. *If no coupling constraints (5.2) were present, the systems would be completely decoupled and could use the (fixed) choice $x_i^s(t) \equiv x_i^{opt}$, where (x_i^{opt}, u_i^{opt}) is the optimal steady-state for system i without coupling constraints as defined in (2.7), i.e., (x_i^{opt}, u_i^{opt}) satisfies $\ell(x_i^{opt}, u_i^{opt}) = \min_{(x_i, u_i) \in \mathcal{S}_i} \ell_i(x_i, u_i)$. In this case, Problem 5.4 would then correspond to Problem 4.23.*

Given the above, a first idea would be that each system uses its current iterate $\xi_i(t)$ of the distributed optimization algorithm as the steady-state $x_i^s(t)$ around which the terminal region in (5.5d) is built (or a projection of $\xi_i(t)$ on the feasible steady-state set in case that the current iterate $\xi_i(t)$ does not satisfy the local input and state constraints or is not a steady-state at all). However, this choice might not be feasible due to the following reasons. Namely, if $|\xi_i(t) - \xi_i(t-1)|$ is large, recursive feasibility of Problem 5.4 might be lost. Furthermore, $x_i^s(t)$ has to be chosen such inside the terminal region around $x_i^s(t)$, the local state and input constraints (5.5c) can be satisfied, which means that (for a given size of the terminal region) equilibria close to the boundary of $\mathbb{Z}_i$ cannot be used. Instead of using $x_i^s(t) := \xi_i(t)$ in (5.5d), in the following we propose a way to gradually change $x_i^s(t)$ "slow enough" such that recursive feasibility of Problem 5.4 can be maintained and we have $\lim_{t \to \infty} x_i^s(t) = \lim_{t \to \infty} \xi_i(t) = x_i^*$. To this end, for each system $i \in \mathbb{I}_{[1,P]}$, let $P_i, Q_i \succ 0$, and define terminal regions of the form

$$\mathbb{X}_i^f(x_i^s, t) := \{x_i \in \mathbb{R}^{n_i} : E_i(x_i, x_i^s) \leq \alpha_i(t)\} \tag{5.7}$$

with $E_i(x_i, x_i^s) := (x_i - x_i^s)^T P_i(x_i - x_i^s)$ and $\alpha_i(t) > 0$ for all $t \in \mathbb{I}_{\geq 0}$. Note that for the (fixed) choice $x_i^s = x_i^*$, this corresponds to the terminal region (4.46) in Section 4.2.1. Now let[2] $\overline{\mathbb{Z}}_i(t) := \mathbb{Z}_i \ominus (\mathbb{X}_i^f(0, t) \times K_i \mathbb{X}_i^f(0, t))$, which we assume to be nonempty for all[3] $t \in \mathbb{I}_{\geq 0}$.

---

[2]For two sets $\mathcal{A}, \mathcal{B} \subseteq \mathbb{R}^n$, the set subtraction $\mathcal{A} \ominus \mathcal{B}$ is defined as $\mathcal{A} \ominus \mathcal{B} := \{x \in \mathbb{R}^n : \{x\} \oplus \mathcal{B} \subseteq \mathcal{A}\}$.

[3]The sets $\overline{\mathbb{Z}}_i(t)$ are nonempty if the set $\mathbb{Z}_i$ has a nonempty interior and the values $\alpha_i(t)$ in (5.7) are chosen small enough. Later, the values $\alpha_i(t)$ are chosen to be nonincreasing in t, and hence the sets $\overline{\mathbb{Z}}_i(t)$ are nonempty if $\alpha_i(0)$ is chosen small enough.

We then impose the following assumption on the terminal regions $\mathbb{X}_i^f$ and the terminal cost V_i^f.

Assumption 5.3. *The terminal regions $\mathbb{X}_i^f$, a local auxiliary controller gain K_i and the continuous terminal cost function $V_i^f(x_i, x_i^s)$ are computed such that the following is satisfied for each state/input equilibrium pair $(x_i^s, u_i^s) \in \mathcal{S}_i \cap \overline{\mathbb{Z}}_i(t)$ and all $x_i \in \mathbb{X}_i^f(x_i^s, t)$:*

i) $(x_i, K_i(x_i - x_i^s) + u_i^s) \in \mathbb{Z}_i$,

ii) $E_i(A_i x_i + B_i(K_i(x_i - x_i^s) + u_i^s), x_i^s) - E_i(x, x_i^s) \leq -(x_i - x_i^s)^T Q_i(x_i - x_i^s)$,

iii) $V_i^f(A_i x_i + B_i(K_i(x_i - x_i^s) + u_i^s), x_i^s) - V_i^f(x_i, x_i^s)$
$\leq -\ell_i(x_i, K_i(x_i - x_i^s) + u_i^s) + \ell(x_i^s, u_i^s)$.

Remark 5.6. *For the (fixed) choice $(x_i^s, u_i^s) = (x_i^*, u_i^*)$, Assumption 5.3 corresponds to Assumption 4.3, with local auxiliary control law $\kappa_i^f(x) =: K_i(x_i - x_i^s) + u_i^s$ and condition ii) slightly strengthened as in Assumption 4.6i). Note that condition i) of Assumption 5.3 is satisfied due to the definition of the set $\overline{\mathbb{Z}}_i(t)$ and the fact that $\mathbb{X}_i^f(x_i^s, t) = \{x_i^s\} \oplus \mathbb{X}_i^f(0, t)$. Furthermore, as discussed in Sections 2.2.1 and 4.2.1 (compare Remark 4.31), in (Amrit et al., 2011) a method was presented how P_i, K_i and V_i^f can be computed such that conditions ii) and iii) of Assumption 5.3 are satisfied with $(x_i^s, u_i^s) = (x_i^*, u_i^*)$. For compact $\mathbb{Z}_i$, this procedure can be modified in a straightforward way to obtain P_i, K_i and V_i^f such that conditions ii) and iii) are satisfied for all $(x_i^s, u_i^s) \in \mathcal{S}_i \cap \overline{\mathbb{Z}}_i(t)$ as required; see (Müller et al., 2014e, Section 4) for more details.*

Now define $c_i := (1 - \frac{\lambda_{\min}(Q_i)}{\lambda_{\max}(P_i)})/\lambda_{\min}(P_i)$, fix $0 < \theta_i < 1$ and let for all $t \in \mathbb{I}_{\geq 0}$

$$\varepsilon_i(t) := \left(-\sqrt{c_i} + \sqrt{c_i + \theta_i \frac{\lambda_{\min}(Q_i)}{\lambda_{\max}(P_i)^2}} \right) \sqrt{\alpha_i(t)}. \tag{5.8}$$

Furthermore, denote by $(\hat{\xi}_i(t), \hat{\eta}_i(t))$ the projection of $(\xi_i(t), \eta_i(t))$ on the set $\mathcal{S}_i \cap \overline{\mathbb{Z}}_i(t)$. We now propose to use the following steady-state $x_i^s(t)$ within the terminal constraint (5.5d):

$$x_i^s(t) := (1 - \lambda_i(t))x_i^s(t-1) + \lambda_i(t)\hat{\xi}_i(t), \tag{5.9}$$

$$\lambda_i(t) := \min \left\{ \frac{\varepsilon_i(t-1)}{|\hat{\xi}_i(t) - x_i^s(t-1)|}, 1 \right\} \tag{5.10}$$

for all $t \in \mathbb{I}_{\geq 1}$ and $x_i^s(0) = x_{i0}^s$, where x_{i0}^s is an arbitrary steady-state satisfying[4] $(x_{i0}^s, u_{i0}^s) \in \mathcal{S}_i \cap \overline{\mathbb{Z}}_i(0)$. Due to convexity of $\mathcal{S}_i \cap \overline{\mathbb{Z}}_i(t)$ and the fact that $\overline{\mathbb{Z}}_i(t+1) \supseteq \overline{\mathbb{Z}}_i(t)$ for all $t \in \mathbb{I}_{\geq 0}$ (the latter will be established below), one can show by induction that $(x_i^s(t), u_i^s(t)) \in \mathcal{S}_i \cap \overline{\mathbb{Z}}_i(t)$ for all $t \in \mathbb{I}_{\geq 0}$, where $u_i^s(t)$ is given by $u_i^s(t) := (1 - \lambda_i(t))u_i^s(t-1) + \lambda_i(t)\hat{\eta}_i(t)$. Furthermore, note that from (5.9)–(5.10) it follows that for all $t \in \mathbb{I}_{\geq 1}$ we have $|x_i^s(t) - x_i^s(t-1)| \leq \varepsilon_i(t-1)$.

It remains to specify how $\alpha_i(t)$ in (5.7), i.e., the size of the terminal region, is updated. Namely, we propose to use the following update rule:

$$\alpha_i(t+1) := \begin{cases} \left(1 - (1 - \theta_i)\frac{\lambda_{\min}(Q_i)}{\lambda_{\max}(P_i)}\right)\alpha_i(t) & \text{if } \lambda_i(t) = 1 \\ \alpha_i(t) & \text{else} \end{cases} \tag{5.11}$$

---

[4]Here, we assume that $\mathcal{S}_i \cap \overline{\mathbb{Z}}_i(0)$ is nonempty, which is the case if there exists at least one state/input equilibrium pair in the interior of $\mathbb{Z}_i$ and $\alpha_i(0)$ in (5.7) is chosen small enough.

for all $t \in \mathbb{I}_{\geq 1}$ and $\alpha_i(1) = \alpha_i(0) := \alpha_{i0} > 0$. Note that from (5.11), it follows that α_i is nonincreasing, as $0 < \theta_i < 1$ and $0 < \lambda_{\min}(Q_i)/\lambda_{\max}(P_i) \leq 1$ (the latter follows from Assumption 5.3(ii), see also the proof of Lemma 5.10 below). This means that the size of the terminal regions is nonincreasing, i.e., for each x_i^s we have $\mathbb{X}_i^f(x_i^s, t+1) \subseteq \mathbb{X}_i^f(x_i^s, t)$ and hence also $\overline{\mathbb{Z}}_i(t+1) \supseteq \overline{\mathbb{Z}}_i(t)$ for all $t \in \mathbb{I}_{\geq 0}$.

Finally, we need to specify the function h_i and the sets $\mathbb{Y}_{i,t}$ appearing in (5.5e). As noted above, the average constraint (5.5e) will be used to ensure convergence of the overall closed-loop system to x^*. Hence, according to the discussion at the end of Section 4.2.2, we use

$$h_i(x_i, u_i, t) = |x_i - x_i^s(t)|, \qquad (5.12)$$

and $\mathbb{Y}_{i,t}$ is defined as in (4.38), i.e.

$$\mathbb{Y}_{i,t+1} := \mathbb{Y}_{i,t} \oplus \mathbb{Y}_i \oplus \overline{\mathbb{Y}}_i(t) \oplus \{-h_i(x_i(t), u_i(t), t)\}, \quad \mathbb{Y}_{i,0} := N\mathbb{Y}_i \oplus \mathbb{Y}_{i00}. \qquad (5.13)$$

The set $\mathbb{Y}_i$ in (5.13) is given as in Assumption 4.10, i.e., $\mathbb{Y}_i = \mathbb{R}_{\leq 0}$, and $\mathbb{Y}_{i00} \subseteq \mathbb{R}$ is again some arbitrary compact set such that (5.5e) is initially feasible. The sets $\overline{\mathbb{Y}}_i(t)$ are defined as

$$\overline{\mathbb{Y}}_i(t) := \overline{y}_i(t)B_1(0), \quad \overline{y}_i(t) := N|x_i^s(t+1) - x_i^s(t)| + |x_i^0(N|t) - x_i^s(t)| \qquad (5.14)$$

for all $t \in \mathbb{I}_{\geq 0}$.

To summarize, the proposed distributed economic MPC scheme is as follows.

Algorithm 5.7 (Distributed economic MPC). *Consider the systems* (5.1). *At each time* $t \in \mathbb{I}_{\geq 0}$, *given the measured states* $x_i(t)$, *all systems* $i \in \mathbb{I}_{[1,P]}$

1. *communicate with neighboring systems,*

2. *perform an iterate of the distributed optimization algorithm solving the minimization problem in* (5.3), *obtaining* $\xi_i(t)$,

3. *solve Problem 5.4, where the terminal region* $\mathbb{X}_i^f$ *in* (5.5d) *is given by* (5.7)–(5.11) *and* h_i *and* $\mathbb{Y}_{i,t}$ *in* (5.5e) *are given by* (5.12)–(5.14),

4. *apply* $u_i(t) := u_i^0(0|t)$.

Algorithm 5.7 is such that each system needs to communicate once in each time step with its neighbors, namely in Step 1. The set of neighboring systems with which system i has to communicate depends on the structure of the coupling constraint set $\mathcal{C}$ and the specific distributed optimization algorithm which is used in Step 2. The latter also determines what information has to be transmitted, such as, e.g., the latest estimate $\zeta_i(t-1)$ or certain dual variables. Furthermore, we remark that within Algorithm 5.7, two optimization problems have to be solved by each system in each time step, one in Step 2 (when performing an iterate of the distributed optimization algorithm) and one in Step 3 when solving Problem 5.4; both Steps 2 and 3 can be performed by all systems in parallel. Finally, note that the optimization problems solved by each system in Step 3 are completely decoupled from each other. Coordination between the systems is achieved via the distributed optimization algorithm (Steps 1 and 2), and information from other systems is injected into Problem 5.4 via the steady-state $x_i^s(t)$ appearing in the constraints (5.5d) and (5.5e).

5.2 Convergence analysis of distributed economic MPC

In this section, we analyze the properties of Algorithm 5.7. To this end, as in Chapter 4, denote by $\mathbb{X}_N(t)$ the set of all states $x \in \mathbb{X}$ such that Problem 5.4 (with $x_i(t) = x_i$) has a solution for all $i \in \mathbb{I}_{[1,P]}$.

Theorem 5.8. *Let $x_0 \in \mathbb{X}_N(0)$ and suppose that Assumptions 5.1–5.3 are satisfied. Then the following is satisfied when applying Algorithm 5.7.*

(i) Problem 5.4 is feasible for all $t \in \mathbb{I}_{\geq 0}$ and all systems $i \in \mathbb{I}_{[1,P]}$.

(ii) $\lim_{t\to\infty}(x_i^s(t), u_i^s(t)) = \lim_{t\to\infty}(\xi_i(t), \eta_i(t)) = (x_i^*, u_i^*)$ for all $i \in \mathbb{I}_{[1,P]}$.

(iii) $\limsup_{T\to\infty} \frac{\sum_{t=0}^T \ell_i(x_i(t), u_i(t))}{T+1} \leq \ell_i(x_i^*, u_i^*)$ for all $i \in \mathbb{I}_{[1,P]}$.

(iv) The overall closed-loop system essentially converges to x^*.

Before proving Theorem 5.8, a couple of remarks and comments are in order. First, due to the fact that $x^* \in \mathcal{C}$, we have the following corollary of Theorem 5.8, ensuring that the coupling constraints are asymptotically satisfied as required.

Corollary 5.9. *Suppose the conditions of Theorem 5.8 are satisfied. Then the closed-loop system essentially converges to the set $\mathcal{C}$.*

Second, we note that as the systems are linear and the cost functions ℓ_i as well as all constraint sets are assumed to be convex, the overall system (including coupling constraints) is optimally operated at the overall optimal steady-state (see Angeli et al. (2009) and Section 2.2). This means that the optimal operational regime for the overall system such that also the coupling constraints are satisfied is operation at the steady-state x^*. Convergence of the overall closed-loop system to x^* is ensured by the (auxiliary) average constraints (5.5e), as shown below in the proof of Theorem 5.8. Without these constraints, the overall closed-loop system might in general not converge to x^*. The reason for this is that the terminal constraint (5.5d) only requires the last state of the predicted state sequence to lie in the terminal region $\mathbb{X}_i^f(x_i^s(t), t)$, and hence even though $\lim_{t\to\infty} x_i^s(t) = x_i^*$ for all $i \in \mathbb{I}_{[1,P]}$, the closed-loop system would in general not converge to x^* due to the use of a general (economic) cost function, and the coupling constraint (5.2) would in general not be satisfied.

Furthermore, in Theorem 5.8 and Corollary 5.9, only essential convergence of the overall closed-loop system to x^* and $\mathcal{C}$, respectively, could be established. The reason for this is that without further assumptions on the convergence rate of $\overline{y}_i$ in (5.14) and hence on the convergence rate of the specific distributed optimization algorithm used in Step 2 of Algorithm 5.7, one can only establish that Assumption 4.9 is satisfied with a possibly unbounded sequence $\hat{\sigma}$ (as shown below in the proof of Theorem 5.8), which results in essential convergence as discussed at the end of Section 4.2.2. On the other hand, if satisfaction of Assumption 4.9 with a bounded sequence $\hat{\sigma}$ can be established (under additional assumptions on the convergence rate of the specific distributed optimization algorithm used in Step 2 of Algorithm 5.7), then asymptotic instead of essential convergence of the overall closed-loop system to x^* follows (compare again the discussion at the end of Section 4.2.2).

Finally, combining statements *(iii)* and *(iv)* of Theorem 5.8, it is straightforward to show that in fact $\lim_{T\to\infty} \frac{\sum_{t=0}^T \ell_i(x_i(t), u_i(t))}{T+1} = \ell_i(x_i^*, u_i^*)$ for each closed-loop system i resulting from

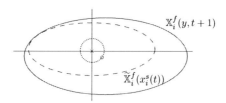

Figure 5.1: Illustration of the sets $\mathbb{X}_i^f(y, t+1)$ (solid), $\widetilde{\mathbb{X}}_i^f(x_i^s(t))$ (dashed) and $\{x_i^s(t)\} \oplus \varepsilon_i(t)B_1(0)$ (dotted), as well as y ("o") and $x_i^s(t)$ ("x").

application of Algorithm 5.7. Hence $\lim_{T\to\infty} \frac{\sum_{t=0}^{T}\sum_{i=1}^{P}\ell_i(x_i(t),u_i(t))}{T+1} = \sum_{i=1}^{P}\ell_i(x_i^*, u_i^*)$, which by optimal steady-state operation is the best achievable asymptotic average performance of the overall system (including coupling constraints). On the other hand, the *transient* average performance of each system can be much better than $\ell_i(x_i^*, u_i^*)$. Namely, one typically observes (compare Section 5.3) that the systems initially "spend time" in a region where the cost ℓ_i is lower than $\ell_i(x_i^*, u_i^*)$ and where the coupling constraints are not satisfied, before they are forced to converge to (x_i^*, u_i^*).

Next, we establish an auxiliary result which will be needed in the proof of Theorem 5.8 later on. It guarantees that the terminal regions $\mathbb{X}_i^f$ defined via (5.7)–(5.11) change "slow enough" such that recursive feasibility of Problem 5.4 can be established.

Lemma 5.10. *Consider the terminal regions $\mathbb{X}_i^f(x_i^s(t), t)$ given by (5.7)–(5.11) and suppose that Assumption 5.3ii) is satisfied. Then for each $t \in \mathbb{I}_{\geq 0}$ and each $x_i \in \mathbb{X}_i^f(x_i^s(t), t)$, it holds that $x_i^+ := A_i x_i + B_i(K_i(x_i - x_i^s(t)) + u_i^s(t)) \in \mathbb{X}_i^f(x_i^s(t+1), t+1)$.*

Proof. As discussed in the paragraph below equation (5.10), we have $(x_i^s(t), u_i^s(t)) \in \mathcal{S}_i \cap \overline{\mathbb{Z}}_i(t)$ for all $t \in \mathbb{I}_{\geq 0}$. Hence from Assumption 5.3ii) with $(x_i^s, u_i^s) = (x_i^s(t), u_i^s(t))$ and using the same calculations as in the proof of Proposition 4.32 (compare (4.48)), for each $t \in \mathbb{I}_{\geq 0}$ and each $x_i \in \mathbb{X}_i^f(x_i^s(t), t)$ we obtain

$$E_i(x_i^+, x_i^s(t)) \leq (1 - \frac{\lambda_{\min}(Q_i)}{\lambda_{\max}(P_i)})\alpha_i(t) = c_i\lambda_{\min}(P_i)\alpha_i(t),$$

where the last equality follows from the definition of c_i in the line above (5.8). Note again that as both Q_i and P_i are positive definite and $E_i \geq 0$, it follows that $0 < \lambda_{\min}(Q_i)/\lambda_{\max}(P_i) \leq 1$. Hence the above means that x_i^+ lies in the tightened set $\widetilde{\mathbb{X}}_i^f(x_i^s(t)) := \{x_i \in \mathbb{R}^{n_i} : E_i(x_i, x_i^s(t)) \leq c_i\lambda_{\min}(P_i)\alpha_i(t)\} \subset \mathbb{X}_i^f(x_i^s(t), t)$. For each $t \in \mathbb{I}_{\geq 0}$, we now want to find the maximum $\varepsilon_i(t) > 0$ such that $\widetilde{\mathbb{X}}_i^f(x_i^s(t)) \subseteq \mathbb{X}_i^f(y, t+1)$ for all $y \in \{x_i^s(t)\} \oplus \varepsilon_i(t)B_1(0)$, where $\mathbb{X}_i^f(y, t+1)$ is defined by (5.7) and (5.11) (see an exemplary illustration in Figure 5.1). This is the case if $E_i(x_i, y) \leq \alpha_i(t+1)$ for all $x_i \in \widetilde{\mathbb{X}}_i^f(x_i^s(t))$ and all $y \in \{x_i^s(t)\} \oplus \varepsilon_i(t)B_1(0)$. We obtain

$$E_i(x_i, y) = (x_i - y)^T P_i(x_i - y)$$
$$= (x_i - x_i^s(t) + x_i^s(t) - y)^T P_i(x_i - x_i^s(t) + x_i^s(t) - y)$$
$$= (x_i - x_i^s(t))^T P_i(x_i - x_i^s(t)) + (x_i^s(t) - y)^T P_i(x_i^s(t) - y) + 2(x_i - x_i^s(t))^T P_i(x_i^s(t) - y)$$
$$\leq c_i\lambda_{\min}(P_i)\alpha_i(t) + \lambda_{\max}(P_i)\varepsilon_i(t)^2 + 2\sqrt{c_i\alpha_i(t)}\lambda_{\max}(P_i)\varepsilon_i(t), \qquad (5.15)$$

where the last inequality follows as $(x_i - x_i^s(t))^T P_i (x_i - x_i^s(t)) \leq c_i \lambda_{\min}(P_i)\alpha_i(t)$ and $|x_i - x_i^s(t)| \leq \sqrt{c_i \alpha_i(t)}$ for all $x_i \in \widetilde{\mathbb{X}}_i^f(x_i^s(t))$, and $|x_i^s(t) - y| \leq \varepsilon_i(t)$ for all $y \in \{x_i^s(t)\} \oplus \varepsilon_i(t)B_1(0)$. By definition of α_i in (5.11), the right hand side of (5.15) is less or equal than $\alpha_i(t+1)$ if we ensure that it is less or equal than $(1 - (1-\theta_i)\frac{\lambda_{\min}(Q_i)}{\lambda_{\max}(P_i)})\alpha_i(t)$. This is the case if $\varepsilon_i(t)$ satisfies the bound (5.8), which follows from solving the corresponding quadratic equation. From (5.9)–(5.10) it follows that for all $t \in \mathbb{I}_{\geq 0}$ we have $|x_i^s(t+1) - x_i^s(t)| \leq \varepsilon_i(t)$, i.e., $x_i^s(t+1) \in \{x_i^s(t)\} \oplus \varepsilon_i(t)B_1(0)$. Hence for each $t \in \mathbb{I}_{\geq 0}$ we have $\widetilde{\mathbb{X}}_i^f(x_i^s(t)) \subseteq \mathbb{X}_i^f(x_i^s(t+1), t+1)$, and therefore $x_i^+ \in \mathbb{X}_i^f(x_i^s(t+1), t+1)$ for each $x_i \in \mathbb{X}_i^f(x_i^s(t), t)$ as claimed. $\square$

Proof of Theorem 5.8. To prove statement *i)* of Theorem 5.8, we show that all conditions for Theorem 4.37*i)* are satisfied[5] for all systems $i \in \mathbb{I}_{[1,P]}$ and hence the conclusion follows. Namely, Assumptions 2.1, 2.3 and 2.6 are satisfied, which follows from (5.1) and Assumptions 5.1 and 5.3. Furthermore, Assumption 4.7 is satisfied by definition of h_i in (5.12) and the fact that $\mathbb{Z}_i$ is compact. Assumption 4.3*i)* is implied by Assumption 5.3*i)* and the fact that $(x_i^s(t), u_i^s(t)) \in \overline{\mathbb{Z}}_i(t)$ for all $t \in \mathbb{I}_{\geq 0}$. Moreover, by Lemma 5.10, we conclude that for the terminal regions given by (5.7)–(5.11), we have that $A_i x_i + B_i(K_i(x_i - x_i^s(t)) + u_i^s(t)) \in \mathbb{X}_i^f(x_i^s(t+1), t+1)$ for each $t \in \mathbb{I}_{\geq 0}$ and each $x_i \in \mathbb{X}_i^f(x_i^s(t), t)$, i.e., Assumption 4.3*ii)* holds. Finally, for each $t \in \mathbb{I}_{\geq 0}$ and each $(x_i, u_i) \in \mathbb{Z}_i$, one obtains

$$h_i(x_i, u_i, t+1) = |x_i - x_i^s(t+1)| = |x_i - x_i^s(t) + x_i^s(t) - x_i^s(t+1)|$$
$$\leq |x_i - x_i^s(t)| + |x_i^s(t) - x_i^s(t+1)| =: h_i(x_i, u_i, t) + \rho(t). \qquad (5.16)$$

Note that (5.16) implies that $h_i(x_i, u_i, t+1) - h_i(x_i, u_i, t) \in \rho(t)B_1(0)$ for each $t \in \mathbb{I}_{\geq 0}$ and each $(x_i, u_i) \in \mathbb{Z}_i$. Moreover, from the definition of $\overline{\mathbb{Y}}_i(t)$ in (5.14), we obtain $\{h_i(x_i^0(N|t), K_i(x_i^0(N|t) - x_i^s(t)) + u_i^s(t), t)\} \oplus N\rho(t)B_1(0) \subseteq \overline{\mathbb{Y}}_i(t) \subseteq \mathbb{Y}_i \oplus \overline{\mathbb{Y}}_i(t)$ for all $t \in \mathbb{I}_{\geq 0}$, which means that the relaxed form of Assumption 4.8 as discussed in Remark 4.38 is satisfied. To summarize, for each system $i \in \mathbb{I}_{[1,P]}$ we can apply Theorem 4.37*i)* (together with Remark 4.38) to conclude that Problem 5.4 is feasible for all $t \in \mathbb{I}_{\geq 0}$.

Next, consider statement *ii)* of Theorem 5.8 and fix $i \in \mathbb{I}_{[1,P]}$. According to (5.11), the sequence $\alpha_i(t)$ is nonincreasing and bounded from below (by zero), hence it converges. Denote its limit by $\alpha_i^{\min} \geq 0$. Then from (5.8) it follows that also $\varepsilon_i(t)$ converges to $\varepsilon_i^{\min}$, where $\varepsilon_i^{\min}$ is given by (5.8) with $\alpha_i(t)$ replaced by $\alpha_i^{\min}$. We now show by contradiction that $\alpha_i^{\min} = 0$. Namely, assume it was not, i.e., $\alpha_i^{\min} > 0$ and hence also $\varepsilon_i^{\min} > 0$. By (5.11), $\alpha_i^{\min} > 0$ is only possible if there exists a finite time $t(\alpha_i^{\min}) \in \mathbb{I}_{\geq 0}$ such that $\lambda_i(t) \neq 1$ for all $t \in \mathbb{I}_{\geq t(\alpha_i^{\min})}$, which means that $\alpha_i(t) = \alpha_i^{\min}$ for all such t. By (5.10), this implies that $|\hat{\xi}_i(t) - x_i^s(t-1)| > \varepsilon_i(t-1) = \varepsilon_i^{\min}$ for all $t \in \mathbb{I}_{\geq t(\alpha_i^{\min})+1}$ and, as $\varepsilon_i(t)$ is nonincreasing, $\lambda_i(t) \geq \lambda_i^{\min} := \varepsilon_i^{\min}/(\max_{r,s \in \mathbb{Z}_i} |r - s|)$ for all $t \in \mathbb{I}_{\geq 1}$. Furthermore, $\mathbb{Z}_i(t) = \mathbb{Z}_i(t(\alpha_i^{\min}))$ for all $t \in \mathbb{I}_{\geq t(\alpha_i^{\min})}$. By Assumption 5.2, $\xi_i(t)$ converges to $x_i^\star$, and hence, thanks to convexity of $\mathcal{S}_i \cap \mathbb{Z}_i(t(\alpha_i^{\min}))$, also $\hat{\xi}_i(t)$ converges to some $\hat{\xi}_i^\alpha$, where $\hat{\xi}_i^\alpha$ is the projection of $x_i^\star$ on the set $\mathcal{S}_i \cap \mathbb{Z}_i(t(\alpha_i^{\min}))$. This means that $\lim_{t \to \infty} \psi_i(t) = 0$, where $\psi_i(t) := \hat{\xi}_i(t) - \hat{\xi}_i^\alpha$. Now

---

[5]Note that Theorem 4.37 was formulated in the setting of one (centralized) system, and hence also the notation used there was slightly different from the one in this chapter (in particular, without the sub-index i in variables, functions etc.). In the following, when verifying that the assumptions of Theorem 4.37 are satisfied, we mean that they are satisfied using the notation of this chapter (e.g., with $\mathbb{Z}$ replaced by $\mathbb{Z}_i$), without explicitly stating this in all places.

consider the sequence $\tilde{x}_i(t) := x_i^s(t) - \hat{\xi}_i^\alpha$. By subtracting $\hat{\xi}_i^\alpha$ on both sides of equation (5.9), we obtain

$$\tilde{x}_i(t) = (1 - \lambda_i(t))\tilde{x}_i(t-1) + \lambda_i(t)\psi_i(t),$$

for all $t \in \mathbb{I}_{\geq 1}$ and hence also

$$|\tilde{x}_i(t)| \leq (1 - \lambda_i^{\min})|\tilde{x}_i(t-1)| + |\psi_i(t)|, \tag{5.17}$$

as $\lambda_i^{\min} \leq \lambda_i(t) \leq 1$ for all $t \in \mathbb{I}_{\geq 1}$ as established above. As $\lim_{t\to\infty} \psi_i(t) = 0$, for each $\delta_\psi > 0$ there exists a $t_\psi \in \mathbb{I}_{\geq 0}$ such that $|\psi_i(t)| \leq \delta_\psi$ for all $t \in \mathbb{I}_{\geq t_\psi}$. Fix $\delta_\psi > 0$ such that $\delta_\psi + 2\delta_\psi/\lambda_i^{\min} \leq \varepsilon_i^{\min}$. From (5.17), it then follows that there exists a finite time $t_{\tilde{x}} \in \mathbb{I}_{\geq t_\psi}$ such that $|\tilde{x}_i(t)| \leq 2\delta_\psi/\lambda_i^{\min}$ for all $t \in \mathbb{I}_{\geq t_{\tilde{x}}}$. Namely, for all $t \in \mathbb{I}_{\geq t_\psi}$, if $|\tilde{x}_i(t)| > 2\delta_\psi/\lambda_i^{\min}$ it follows from (5.17) that $|\tilde{x}_i(t+1)| - |\tilde{x}_i(t)| \leq -\delta_\psi$, and hence $|\tilde{x}_i(t_{\tilde{x}})| \leq 2\delta_\psi/\lambda_i^{\min}$ for some finite time $t_{\tilde{x}} \in \mathbb{I}_{\geq t_\psi}$. But then, from (5.17) it follows that also $|\tilde{x}_i(t)| \leq 2\delta_\psi/\lambda_i^{\min}$ for all $t \in \mathbb{I}_{\geq t_{\tilde{x}}}$. Summarizing the above, there exists a time $t' \in \mathbb{I}_{\geq t(\alpha_i^{\min})}$ such that

$$
\begin{aligned}
|\hat{\xi}_i(t') - x_i^s(t'-1)| &= |\psi_i(t') - \tilde{x}_i(t'-1)| \\
&\leq |\psi_i(t')| + |\tilde{x}_i(t'-1)| \leq \delta_\psi + 2\delta_\psi/\lambda_i^{\min} \leq \varepsilon_i^{\min},
\end{aligned}
$$

which is a contradiction to the fact established above that $|\hat{\xi}_i(t) - x_i^s(t-1)| > \varepsilon_i^{\min}$ for all $t \in \mathbb{I}_{\geq t(\alpha_i^{\min})}$. Hence we conclude that $\alpha_i^{\min} = 0$. But then $\lim_{t\to\infty} \mathcal{S}_i \cap \overline{\mathbb{Z}}_i(t) = \mathcal{S}_i$ and hence $\lim_{t\to\infty}(\hat{\xi}_i(t), \hat{\eta}_i(t)) = \lim_{t\to\infty}(\xi_i(t), \eta_i(t)) = (x_i^*, u_i^*)$ due to Assumption 5.2 and the definition of $(\hat{\xi}_i, \hat{\eta}_i)$. Furthermore, by (5.11), there exists an infinite subsequence $\{t_j\} \subseteq \mathbb{I}_{\geq 0}$ such that $\lambda(t_j) = 1$ and hence $x_i^s(t_j) = \hat{\xi}_i(t_j)$. As $\hat{\xi}_i(t)$ converges to x_i^*, for each $\delta > 0$ there exists a $t(\delta) \in \mathbb{I}_{\geq 0}$ such that $|\hat{\xi}_i(t) - x_i^*| \leq \delta$ for all $t \in \mathbb{I}_{\geq t(\delta)}$. Let $j(\delta) := \min_{t_j \geq t(\delta)} j$; then it follows from (5.9) that also $|x_i^s(t) - x_i^*| \leq \delta$ for all $t \in \mathbb{I}_{\geq t_{j(\delta)}}$. This can be shown by induction as $|x_i^s(t_{j(\delta)}) - x_i^*| \leq \delta$ and from (5.9) it follows that if both $|x_i^s(t-1) - x_i^*| \leq \delta$ and $|\hat{\xi}_i(t) - x_i^*| \leq \delta$ for some t, then also the convex combination $x_i^s(t)$ satisfies $|x_i^s(t) - x_i^*| \leq \delta$. But as $\delta > 0$ was arbitrary, it follows that $\lim_{t\to\infty} x_i^s(t) = x_i^*$, as claimed. In the same way, it follows that $\lim_{t\to\infty} u_i^s(t) = u_i^*$.

To prove statement *(iii)*, fix again $i \in \mathbb{I}_{[1,P]}$ and consider two consecutive time instants $t \in \mathbb{I}_{\geq 0}$ and $t+1$. Using again the candidate input sequence $\tilde{\boldsymbol{u}}_i(t+1) := \{u_i^0(1|t), \ldots, u_i^0(N-1|t), u_i'\}$ with $u_i' := K_i(x_i(N|t) - x_i^s(t)) + u_i^s(t)$ and the corresponding candidate state sequence $\tilde{\boldsymbol{x}}_i(t+1) := \{x_i^0(1|t), \ldots, x_i^0(N|t), A_i x_i^0(N|t) + B_i u_i'\}$, we obtain

$$
\begin{aligned}
J_{i,N}^0(x(t+1)) - J_{i,N}^0(x(t)) &\leq -\ell_i(x_i(t), u_i(t)) + \ell_i(x_i^0(N|t), u_i') \\
&\quad + V_i^f(A_i x_i^0(N|t) + B_i u_i', x_i^s(t+1)) - V_i^f(x_i^0(N|t), x_i^s(t)) \\
&= -\ell_i(x_i(t), u_i(t)) + \ell_i(x_i^0(N|t), u_i') + \nu_i(t+1) \\
&\quad + V_i^f(A_i x_i^0(N|t) + B_i u_i', x_i^s(t)) - V_i^f(x_i^0(N|t), x_i^s(t)) \\
&\overset{\text{Ass. 5.3}iii}{\leq} -\ell_i(x_i(t), u_i(t)) + \ell_i(x_i^s(t), u_i^s(t)) + \nu_i(t+1) \\
&= -\ell_i(x_i(t), u_i(t)) + \ell_i(x_i^*, u_i^*) + \mu_i(t+1) + \nu_i(t+1)
\end{aligned}
$$

where $\nu_i(t+1) := V_i^f(A_i x_i^0(N|t) + B_i u_i', x_i^s(t+1)) - V_i^f(A_i x_i^0(N|t) + B_i u_i', x_i^s(t))$ and $\mu_i(t+1) := \ell_i(x_i^s(t), u_i^s(t)) - \ell_i(x_i^*, u_i^*)$. By continuity of ℓ and V_f and the fact that

$\lim_{t\to\infty}(x_i^s(t), u_i^s(t)) = (x_i^*, u_i^*)$ by statement $ii)$ of the theorem, we obtain $\lim_{t\to\infty}\mu_i(t) = \lim_{t\to\infty}\nu_i(t) = 0$. Hence, using similar calculations as in the proof of Theorem 4.65, we conclude that for the closed-loop system i resulting from application of Algorithm 5.7, we have $\limsup_{T\to\infty}\frac{\sum_{t=0}^{T}\ell_i(x_i(t),u_i(t))}{T+1} \leq \ell_i(x_i^*, u_i^*)$ as claimed.

Finally, we prove statement $iv)$ of Theorem 5.8, which will be done by applying Theorem 4.37$iii)$; hence we have to show that Assumption 4.9 is satisfied for all $i \in \mathbb{I}_{[1,P]}$. To this end, note that

$$h(x_i, u_i, t) = |x_i - x_i^s(t)| = |x_i - x_i^* + x_i^* - x_i^s(t)|$$
$$\leq |x_i - x_i^*| + |x_i^* - x_i^s(t)| =: |x_i - x_i^*| + \hat{\varphi}(t).$$

Defining $\hat{\sigma}(t) := \sum_{k=0}^{t-1}\left(\overline{y}_i(k) + \hat{\varphi}(k)\right)$ with $\overline{y}_i(t)$ as in (5.14), it follows that $\sum_{k=0}^{t-1}\left(\overline{Y}_i(k) \oplus \hat{\varphi}(k)B_1(0)\right) = \hat{\sigma}(t)B_1(0)$ for all $t \in \mathbb{I}_{\geq 0}$. Thus, in order to apply Theorem 4.37$iii)$ with $\hat{h}_i(x_i, u_i) := |x_i - x_i^*|$, it remains to show that $\lim_{t\to\infty}\hat{\sigma}(t)/t = 0$. To this end, note that for all $t \in \mathbb{I}_{\geq 0}$, from (5.9)–(5.10) it follows that $|x_i^s(t+1) - x_i^s(t)| \leq \varepsilon_i(t)$; furthermore, the definition of $\mathbb{X}_i^f$ in (5.7) together with the terminal constraint (5.5d) implies that $|x_i^0(N|t) - x_i^s(t)| \leq \sqrt{\alpha_i(t)/\lambda_{\min}(P_i)}$. According to (5.14), this means that $\overline{y}_i(t) \leq N\varepsilon_i(t) + \sqrt{\alpha_i(t)/\lambda_{\min}(P_i)}$. As proven in statement (ii) of the Theorem, $\lim_{t\to\infty}\alpha_i(t) = \lim_{t\to\infty}\varepsilon_i(t) = 0$, and hence also $\lim_{t\to\infty}\overline{y}_i(t) = 0$. Furthermore, $\lim_{t\to\infty}x_i^s(t) = x_i^*$ and hence $\lim_{t\to\infty}\hat{\varphi}(t) = 0$. But this means that also $\lim_{t\to\infty}\overline{y}_i(t) + \hat{\varphi}(t) = 0$, and hence $\lim_{t\to\infty}\hat{\sigma}(t)/t = 0$. Thus we can apply Theorem 4.37$iii)$ with $\hat{h}_i(x_i, u_i) := |x_i - x_i^*|$ to conclude that for the closed-loop system i resulting from application of Algorithm 5.7 we have $Av[\hat{h}_i(\boldsymbol{x}_i, \boldsymbol{u}_i)] \subseteq \mathbb{Y}_i$, which translates into

$$\lim_{T\to\infty}\frac{\sum_{t=0}^{T}|x_i(t) - x_i^*|}{T+1} = 0.$$

We can now apply Lemma 1 (together with Remark 13) from Müller et al. (2014f) (see also Lemma 2.2 in Angeli et al. (2011)) to conclude that the closed-loop system i essentially converges to x_i^* (compare the discussion at the end of Section 4.2.2). As this holds for all systems $i \in \mathbb{I}_{[1,P]}$, it follows that the overall closed-loop system resulting from application of Algorithm 5.7 essentially converges to x^* as claimed. $\qquad\square$

Remark 5.11. *The proof of Theorem 5.8 can be modified such that Assumption 5.1 can be relaxed to allow for possibly unbounded constraint sets $\mathbb{Z}_i$ (again under the assumption that S^* is well defined, i.e., the minimum in (5.3) exists). In this case, additional conditions on the stage and terminal cost functions ℓ_i and V_i^f are needed similar to what was discussed in the paragraph below Theorem 2.5. Furthermore, in this case also h_i as defined in (5.12) is possibly unbounded on $\mathbb{Z}_i \times \mathbb{I}_{\geq 0}$ and hence Assumption 4.7 is not satisfied. This assumption was needed in Theorem 4.37 (which was used in the proof of Theorem 5.8) to conclude that (4.52) follows from (4.51). But this conclusion is in fact also true without boundedness of h_i for the special choice of $\mathbb{Y}_i$ and h_i used in this chapter, i.e., $\mathbb{Y}_i = \mathbb{R}_{\leq 0}$ and h_i given by (5.12). Finally, we note that the procedure discussed in Remark 5.6 how to calculate the terminal ingredients is only valid if $\lambda_{\max}(\nabla_x^2\overline{\ell}(x)) \leq c$ for some $c \in \mathbb{R}$, all $(x_i^s, u_i^s) \in \mathcal{S}_i$ and all $x \in \mathbb{X}_i^f(x_i^s, 0)$, where $\overline{\ell}(x) := \ell_i(x_i, K_i(x_i - x_i^s) + u_i^s) - \ell_i(x_i^s, u_i^s)$. This needs to be assumed in case that $\mathbb{Z}_i$ is unbounded and is, e.g., satisfied if ℓ is linear or quadratic.*

Remark 5.12. *As can be seen from (5.9)–(5.10), the convergence rate of x_i^s depends on the convergence rate of ξ_i, i.e., on the convergence rate of the specific distributed optimization algorithm used in Step 2 of Algorithm 5.7, and also on ε_i. Concerning the latter, the value of ε_i given by (5.8) in general only yields (via (5.10)) a lower bound on how fast x_i^s can change such that recursive feasibility of Problem 5.4 can be ensured. Instead, at each time $t \in \mathbb{I}_{\geq 1}$, between Steps 2 and 3 of Algorithm 5.7, each system could also solve the additional optimization problem*

$$\max_{\lambda_i(t) \leq 1, (5.5a) - (5.5d), (5.7), (5.9)} \lambda_i(t) \tag{5.18}$$

and then, in Step 3 of Algorithm 5.7, use the optimal value $\lambda_i^*(t)$ in (5.9) instead of $\lambda_i(t)$ given by (5.10) (or any $\lambda_i(t)$ satisfying $0 \leq \lambda_i(t) \leq \lambda_i^*(t)$), in order to increase the convergence speed of x_i^s. Furthermore, in the special case where all steady-states $x \in \mathcal{S}_{\mathbb{X}}$ are reachable in N steps from all feasible initial conditions $x(0) \in \mathbb{X}_N(0)$ (which is, e.g., the case in the example in Section 5.3), one can use $\lambda_i^*(t) = 1$ for all $t \in \mathbb{I}_{\geq 0}$ without having to solve the additional optimization problem (5.18). Finally, as discussed above, convergence of x_i is ensured by the constraint (5.5e), and the convergence speed depends on the set $\mathbb{Y}_{i00}$ and the convergence rate of $\bar{y}_i$ (compare Remarks 4.25 and 4.38), where the latter, in turn, depends on the convergence rate of x_i^s according to (5.14).

5.3 Application: Consensus under conflicting objective

We now illustrate Algorithm 5.7 with the problem of reaching consensus under conflicting objectives. To this end, consider eight two-dimensional discrete-time double integrator systems of the form (5.1) with $x_i = [x_{1i} \ x_{2i}]^T \in \mathbb{R}^2$ and $u_i \in \mathbb{R}$. For all $i \in \mathbb{I}_{[1,8]}$, the system matrices are given by $A_i = \begin{bmatrix} 1 & 1 \\ 0 & 1 \end{bmatrix}$ and $B_i = \begin{bmatrix} 0 & 1 \end{bmatrix}^T$, respectively, and the input and state constraint sets by $\mathbb{Z}_i = \mathbb{X}_i \times \mathbb{U}_i = [-5,5]^3$. The set of all feasible state/input equilibrium pairs is $\mathcal{S}_i = \{x_{1i} \in [-5,5], x_{2i} = u_i = 0\}$, and the self-interest of each system is modeled through the cost function $\ell_i(x_i, u_i) = (x_{1i} - a_i)^2 + (x_{2i} - b_i)^2 + (u_i - d_i)^2$, where a_i, b_i, d_i are randomly chosen within the interval $[-3,3]$. The cooperative requirement the systems have to fulfill is to asymptotically reach consensus, i.e., $x_1 = \cdots = x_8$ asymptotically. As discussed in Section 3.3, this translates into a coupling constraint (5.2) given by $(E(\mathcal{G})^T \otimes I_2)x = 0$, where $E(\mathcal{G})$ is the incidence matrix of the graph $\mathcal{G}$ describing the interconnection topology of the systems, which we choose to be a circle graph. Without the coupling constraint (5.2), the optimal steady-state for each system would be given by $x_{1i}^{\text{opt}} = a_i$ and $x_{2i}^{\text{opt}} = u_i^{\text{opt}} = 0$. In case that the values a_i are not the same for all systems, the individual objectives of the systems (given by ℓ_i) are conflicting with the requirement of reaching consensus, as $x^{\text{opt}} := [(x_1^{\text{opt}})^T \ \cdots \ (x_5^{\text{opt}})^T]^T \notin \mathcal{C}$. Figure 5.2 shows simulation results obtained by applying Algorithm 5.7, where in Step 2 a distributed dual subgradient algorithm is used. The prediction horizon in Problem 5.4 is $N = 10$, and the overall optimal state/input equilibrium pair according to (5.3) is such that $x_{1i}^* = -0.612$, $x_{2i}^* = 0$ and $u_i^* = 0$ for all $i \in \mathbb{I}_{[1,8]}$. Both ξ_i and x_i^s are initialized with $\xi_i(0) = x_i^s(0) = x_i^{\text{opt}}$, and the initial conditions $x_i(0)$ are randomly chosen within the set $[-3,3] \times [-2,2]$. As discussed in the second point of Remark 4.38, the sets $\mathbb{Y}_{i,t}$ on the right hand side of (5.13) are replaced by $\tilde{\mathbb{Y}}_{i,t} = \max\{\sum_{k=0}^{N-1} h_i(x_i^0(k|t), u_i^0(k|t)), \min\{\mathbb{Y}_{i,t}, \mathbb{Y}_{i,0}\}\}$ in order to avoid that the sets $\mathbb{Y}_{i,t}$ become unnecessarily large, which would result in a slow convergence of x_i.

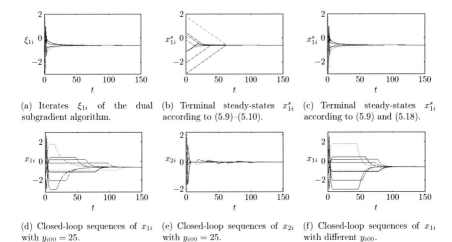

(a) Iterates ξ_{1i} of the dual subgradient algorithm.

(b) Terminal steady-states x_{1i}^s according to (5.9)–(5.10).

(c) Terminal steady-states x_{1i}^s according to (5.9) and (5.18).

(d) Closed-loop sequences of x_{1i} with $y_{i00} = 25$.

(e) Closed-loop sequences of x_{2i} with $y_{i00} = 25$.

(f) Closed-loop sequences of x_{1i} with different y_{i00}.

Figure 5.2: Closed-loop sequences for the example in Section 5.3.

Figure 5.2 shows that as guaranteed by Theorem 5.8, ξ_i, x_i^s and x_i converge to x_i^*. In Figure 5.2(b), $x_i^s(t)$ is calculated according to (5.9)–(5.10); during the initial phase, one observes that $\lambda_i(t) < 1$ and hence $|x_i^s(t) - x_i^s(t+1)| = \varepsilon_i(t-1)$. On the other hand, Figure 5.2(c) shows $x_i^s(t)$ calculated according to (5.9) and (5.18). In fact, one obtains that $\lambda_i(t) = 1$ and hence $x_i^s(t) = \hat{\xi}_i(t)$ for all $t \in \mathbb{I}_{\geq 0}$, which is the case as all steady-states $x \in \mathcal{S}_{\mathbb{X}}$ are reachable in N steps from all feasible initial conditions $x(0) \in \mathbb{X}_N(0)$. Hence one can use $\lambda_i \equiv 1$ without having to solve the additional optimization problem (5.18), as pointed out in Remark 5.12. Figures 5.2(d)–5.2(f) show closed-loop state sequences. One observes that as discussed in Section 5.2, during the transient phase each system "spends time" in a region where its cost ℓ_i is lower than $\ell(x_i^*, u_i^*)$, and in particular approaches $x_i^{\text{opt}} = [a_i\ 0]^T$, before it is forced to converge to x_i^* in order to asymptotically reach consensus. As discussed in Section 4.2 (compare Remark 4.25), the length of this transient phase can be adjusted by using the sets $\mathbb{Y}_{i00}$ in (5.13) as tuning parameters, where a larger $\mathbb{Y}_{i00}$ yields a longer transient phase. In Figures 5.2(d)–5.2(e), $\mathbb{Y}_{i00} = \{y \in \mathbb{R} : |y| \leq y_{i00}\}$ with $y_{i00} = 25$ is used for all systems $i \in \mathbb{I}_{\geq 0}$, which results in the fact that those systems for which $|x_i^{\text{opt}} - x_i^*|$ is large are forced to converge faster than those systems for which $|x_i^{\text{opt}} - x_i^*|$ is smaller. On the other hand, in Figure 5.2(f), we use $y_{100} = y_{800} = 35$, $y_{200} = y_{400} = 15$, $y_{300} = y_{600} = 80$, $y_{500} = 50$, and $y_{800} = 25$.

5.4 Summary and discussion

In this chapter, we have presented a distributed economic MPC scheme for the cooperative control of systems with potentially conflicting individual objectives. The proposed method exhibits a two-layer structure, where the first layer consists of a distributed optimization algorithm determining the overall optimal state/input equilibrium pair, whose iterates are then used within the economic MPC problems for each system in the second layer. We showed that the overall closed-loop system converges to the overall optimal steady-state x^* and hence also to the set $\mathcal{C}$, i.e., the cooperative control task is satisfied as required. On the other hand, during the transient phase each of the systems can optimize its individual performance, i.e., act according to its self-interest as desired. Hence in conclusion, we find that (distributed) economic MPC is a tool which is well suited for the cooperative control of self-interested systems.

One of the key advantages of the proposed distributed economic MPC scheme is that the overall optimal state/input equilibrium pair does not have to be known a priori, but is determined online. This makes the method readily applicable in a plug-and-play setting (Stoustrup (2009); compare also Riverso et al. (2013)), i.e., where systems leave or new systems enter the network. Namely, if such an event occurs, the entering or leaving system has to notify its neighbors in order to set up or terminate communication, and then the systems just keep running Algorithm 5.7, which means that the possibly changed overall optimal state/input equilibrium pair is again determined online and the results of Theorem 5.8 are still valid.

In this chapter, the cooperative requirement which the systems have to fulfill was translated into the asymptotic fulfillment of the coupling constraints (5.2). While this setting, as discussed above, includes a variety of cooperative control tasks such as output agreement problems, other cooperative control objectives can rather be formulated as satisfaction of some coupling constraints at *all* time instants instead (or as a combination of both). For example, this can be the case in distributed power generation problems (see, e.g., Wood and Wollenberg (1996), and also Spudić and Baotić (2013)), where the cooperative task is to meet the overall required power production all time instants. In such cases, Algorithm 5.7 needs to be modified such that the coupling constraints are guaranteed to be satisfied at all time instants, which is a subject for future research. One approach could be to use the techniques developed in Chapter 3, i.e., the systems solve Problem 5.4 in Step 3 of Algorithm 5.7 in a sequential order. In this case, additional communication between the systems is needed within Step 3, i.e., predicted state sequences have to be transmitted as discussed in Chapter 3.

Finally, we note that it would be interesting to extend the results presented in this chapter to nonlinear systems with nonconvex cost functions and constraint sets. However, several difficulties arise in this respect. First, it is in general difficult to satisfy Assumption 5.2 if the minimization problem in (5.3) is nonconvex[6]. Furthermore, x_i^s as determined by (5.9) is in general not a steady-state if the systems are nonlinear, and hence the gradual change of the terminal regions such that recursive feasibility can be ensured has to be done in a different way. Finally, in case that the overall system is not dissipative with respect to the supply rate $s(x, u) = \sum_{i=1}^P \ell_i(x_i, u_i) - \ell(x_i^*, u_i^*)$, it is in general not optimally operated at steady-state (compare Section 4.1.1), which means that enforcing convergence to x^* is not the optimal operational regime. In this case, one might rather think of different strategies of how to satisfy the coupling constraints (i.e., the cooperative control task), involving, for example, again the exchange of predicted state sequences at each time instant.

---

[6]On the other hand, we note that in nonlinear (centralized) economic MPC, one typically also assumes that the optimal steady-state can be computed.

Chapter 6

Conclusions

We now summarize the main results of this thesis and discuss possible topics for future research.

6.1 Summary and discussion

In this thesis, various results in the fields of economic and distributed model predictive control have been obtained. In both fields, it is of interest to go beyond the classical control objective of (setpoint) stabilization, but rather to consider the optimization of some general, possibly economic, performance criterion and the achievement of some general cooperative control task, respectively. The results in this thesis contribute to a deepened systems theoretic understanding of the phenomena encountered in these type of control problems and provide various methods for their solution.

In particular, in Chapters 3 and 5, we developed two frameworks for using distributed MPC in cooperative control problems, where a network of (dynamically decoupled) systems has to achieve some common, cooperative goal. The first one (Chapter 3) was formulated in the context of stabilizing MPC, while the second (Chapter 5) was expressed in the setting of economic MPC. In particular, the former framework is suited for situations where the systems use (coupled) cost functions which are consistent with the cooperative goal to be achieved. We presented two distributed MPC schemes which ensure satisfaction of coupling constraints between the systems and achievement of the cooperative goal, one for general and one for separable cost functions. Both schemes are applicable to general cooperative control problems; as a specific example, we showed how they can be used for consensus and synchronization problems. On the other hand, the framework developed in Chapter 5 considers the situation where each system seeks to optimize its own, individual objective, i.e., act according to its self-interest, which might not be consistent with the cooperative goal to be achieved; this naturally results in an economic MPC setting. The proposed control structure consists of two layers, where the first ensures cooperation between the systems, which is done by determining the overall optimal steady-state using some distributed optimization algorithm. In the second layer, each system computes its control actions by solving an economic MPC problem involving its own, individual objective, and using information from the upper layer in order to achieve the cooperative goal. In conclusion, the results presented in Chapters 3 and 5 provide a variety of tools for the solution of cooperative control problems, showing that distributed MPC is a method which is well suited for such control tasks.

In Chapter 4, we first thoroughly investigated a dissipativity condition which is crucial in economic MPC for classifying optimal steady-state operation and for establishing stability. We showed that under a rather mild controllability condition, dissipativity is not only sufficient, but in fact also necessary for optimal steady-state operation, which means that dissipativity is a precise characterization (and not only a possibly conservative sufficient condition) for this property. Furthermore, these results are also important for stability analysis in economic MPC. Namely, based on certain dynamic properties of the system, they ensure existence of a storage function λ such that the considered dissipativity condition is satisfied, which (strengthened to strict dissipativity) is in turn sufficient for establishing stability. Furthermore, we analyzed robustness of the dissipativity condition with respect to changes in the constraint set. This is important as the supply rate depends on the optimal steady-state and hence the constraints. Under certain regularity conditions, we established robustness of dissipativity with respect to small and large changes in the constraint set, the latter under an additional convexity assumption. We note that these results can also be used when considering networks of systems as in Chapters 3 and 5, for example in order to determine whether the overall system (including coupling constraints) is dissipative given that each single system is dissipative.

The second contribution in Chapter 4 was to study economic MPC schemes including average constraints. We showed how an appropriate economic MPC scheme can be formulated such that both transient and asymptotic average constraints are guaranteed to be satisfied for the resulting closed-loop system. Furthermore, in case that the system is strictly dissipative and hence optimally operated at steady-state, we provided a Lyapunov-like analysis to show that the closed-loop system asymptotically converges to the optimal steady-state. Notably, the employed Lyapunov function is different than the one typically used in economic MPC without average constraints. As discussed earlier, average constraints are of particular interest in economic MPC as the closed-loop system is not necessarily convergent, and hence the results presented in Section 4.2 can be seen as an important tool in order to include these constraints into the design of economic model predictive controllers.

Finally, the third contribution in Chapter 4 was to develop an economic MPC scheme with a self-tuning terminal cost. In particular, we showed how the terminal cost can be defined in a self-tuning fashion such that desirable bounds on the closed-loop asymptotic average performance can be guaranteed. The proposed scheme uses a generalized terminal constraint and hence is well suited in applications where the optimal steady-state is unknown a priori, or where a comparatively short prediction horizon is desired which would result in a possibly very small feasible set when using a standard (fixed) terminal constraint.

In summary, the results presented in this thesis provide novel insight and solution methods to various problems encountered in the fields of distributed and economic MPC, and hence contribute to the possibility to apply model predictive control in situations where the objective lies beyond setpoint stabilization.

6.2 Outlook

The results obtained in this thesis can serve as a starting point for various interesting future research topics, as outlined in the following.

In economic MPC, as discussed in Section 4.1, the examined dissipativity condition is used for classifying optimal steady-state operation of a system and for establishing stability of the optimal steady-state. If this dissipativity condition is not satisfied, then the system is in general not optimally operated at steady-state, as shown in Section 4.1, but some non-constant solution exists which results in a better asymptotic average performance. Hence it would be interesting to extend suitable dissipativity conditions to the case where periodic operation of the system is optimal. Some first results in this direction have recently been obtained by (Grüne and Zanon, 2014; Zanon et al., 2013), where different sufficient conditions for optimal periodic operation of a system were identified. The results of Section 4.1 provide the basis for a thorough analysis of these dissipativity conditions, in particular with respect to necessity for optimal periodic operation as well as robustness issues with respect to changes in the constraints. Furthermore, as already discussed in Section 4.1, an important open problem is to examine necessity of dissipativity in case of systems subject to average constraints, where a relaxed dissipativity condition as used in Section 4.2 is of interest.

The economic MPC schemes developed in Section 4.2, which ensure satisfaction of both transient and asymptotic average constraints, were obtained using a terminal region around the (fixed) optimal steady-state. An interesting topic for future research is to investigate how these MPC schemes can be modified such that fulfillment of transient and asymptotic average constraints can also be guaranteed when using a generalized terminal constraint (such as in Section 4.3) or when using no terminal constraints at all (such as in (Grüne, 2013)). In this respect, one in particular has to examine how the optimization problems can be formulated such that recursive feasibility of the additional constraints, which are needed to guarantee satisfaction of the average constraints for the closed-loop system, can be ensured.

In Chapters 4 and 5, asymptotic average performance bounds were obtained for various different economic MPC schemes. However, no theoretical guarantees for the *transient* average performance of the systems, i.e., the closed-loop performance averaged over some finite time interval, could be given. The same is true for the vast majority of the available economic MPC literature. Some first results concerning transient performance were obtained in (Grüne, 2013; Grüne and Stieler, 2014) for an economic MPC setting without terminal constraints and where the optimal steady-state is (practically) asymptotically stable for the closed-loop system. It would be worthwhile to establish similar transient average performance bounds for economic MPC with terminal constraints, and also for economic MPC with transient and asymptotic average constraints. Concerning the latter, it would be of particular interest to quantify the effect of the tuning parameter $\mathbb{Y}_{00}$ on the achievable transient performance (compare Remark 4.25).

The results in this thesis were obtained for nominal system dynamics only, i.e., systems without disturbances were considered. For standard tracking MPC, many different methods such as tube-based MPC, min-max MPC or input-to-state stable MPC exist in order to deal with uncertainties and disturbances, see, e.g., (Limon et al., 2009; Mayne et al., 2005; Raimondo et al., 2009; Raković et al., 2012; Yu, 2011) and the references therein. On the other hand, only very little research has been done in this direction in the context of economic MPC. Huang et al. (2012) consider robustness issues of an economically oriented MPC scheme, which, however, uses a (standard) tracking cost function. Furthermore, Hovgaard et al. (2011) present an application case study of economic MPC under uncertainties. As the main interest in economic MPC is the optimization of the closed-loop performance, it is crucial to ensure that the disturbances do not result in a possibly large performance degradation. However, this might occur when just transferring existing robust MPC methods from a tracking to an economic MPC context, as demonstrated in a recent paper by Bayer et al. (2014). In order to overcome this problem, the tube-based economic MPC scheme developed in this reference takes the disturbances explicitly into account within the cost function. Considering the above, identifying and developing suitable robust economic MPC frameworks to which also the results of this thesis can be extended is another interesting subject for future research.

The distributed MPC schemes in Chapters 3 and 5 were developed under the assumption that there are no failures in the communication between the systems, such as delays or packet losses. In real applications, such network deficiencies are always present, and hence studying the impact of these network-induced effects on the distributed MPC schemes presented in this thesis is an important future research topic. A particular challenge in this respect is how to ensure satisfaction of coupling constraints despite the fact that each system might not know the newest predictions of its neighbors. A first step in this direction has recently been obtained by (Schaich, 2013; Schaich et al., 2014), where the distributed MPC schemes developed in Chapter 3 were studied in networks with packet losses for the special control task of setpoint stabilization. Furthermore, besides considering communication imperfections between different systems, one would also need to consider additional network-induced effects if also each system itself is connected to its controller over a network. A good starting point in this respect could be the results developed in a centralized setting, i.e., for one system and controller, such as, e.g., (Findeisen et al., 2011) and (Grüne et al., 2014, Section 4.2). Moreover, in various cooperative control tasks it makes sense to consider a general time-varying or "switching" communication topology (see, e.g., Olfati-Saber et al. (2007)), which is for example the case when each system can only communicate with neighboring systems within a limited sensing radius. Again, a crucial question is how coupling constraints can be satisfied in this case, and, if coupled cost functions as in Chapter 3 are used, how a decaying of the overall optimal value function can be achieved. To this end, tools from the switched systems literature such as multiple Lyapunov functions might be useful, see, e.g., (Branicky, 1998; Liberzon, 2003). Again, a first step towards a distributed MPC scheme for networks with switching topology was recently obtained by Schaich (2013) for the special control task of setpoint stabilization.

The distributed MPC schemes presented in Chapter 3 as well as the needed assumptions were formulated such that they are suitable for general cooperative control tasks; as a specific application, we then showed how the terminal ingredients can be calculated in consensus and synchronization problems such that the required assumptions are satisfied. Hence another topic of future research is to develop procedures how the terminal ingredients can be calculated for different specific cooperative control objectives. Furthermore, while the MPC schemes presented in Chapter 3 are truly distributed in the sense that only neighboring communication is needed, the offline design procedure for the terminal ingredients in consensus and synchronization problems developed in Section 3.3 requires that a centralized LMI is solved. While this can be acceptable in some applications, in others one rather wants also the offline design phase to be distributed. Hence it is also of interest to develop such distributed design procedures, similar to what was done for the control objective of setpoint stabilization in Conte et al. (2012).

Finally, as already discussed in Section 5.4, there are various interesting future research directions in the field of distributed economic MPC. Namely, it would be interesting to combine ideas from Chapters 3 and 5 in order to develop a distributed economic MPC framework for cooperative control problems which possibly involve satisfaction of coupling constraints at all time instants. Furthermore, the distributed economic MPC scheme in Chapter 5 was developed for linear systems with convex constraints and cost functions only. Hence it would be very interesting to explore possible extensions to nonlinear systems and nonconvex constraints and cost functions. Besides some technical challenges, the main conceptual difference compared to the linear case is that the overall system is not necessarily optimally operated at steady-state, and hence new ways how to ensure cooperation and coordination between the systems have to be found. For more details on these issues, we refer the reader to Section 5.4.

In conclusion, the results presented in this thesis open up various research directions, ranging from a further theoretical analysis of economic MPC schemes to their application in distributed cooperative control problems, which constitute important steps on the way beyond setpoint stabilization in model predictive control.

Appendix A

Stability of discrete-time systems

A.1 Stability and asymptotic stability

Consider a discrete-time nonlinear system of the form

$$x(t+1) = f(x(t)), \qquad x(0) = x_0, \tag{A.1}$$

with $t \in \mathbb{I}_{\geq 0}$, where $x(t) \in \mathbb{R}^n$ is the system state at time t and $x_0 \in \mathbb{R}^n$ is the initial condition. A closed set $\mathcal{A} \subseteq \mathbb{R}^n$ is *positively invariant* for system (A.1) if $f(x) \in \mathcal{A}$ for all $x \in \mathcal{A}$. In this thesis, we use the following standard definitions of stability and asymptotic stability (see, e.g., Jiang and Wang (2002); Rawlings and Mayne (2009)).

Definition A.1. *A closed and positively invariant set $\mathcal{A} \subseteq \mathbb{R}^n$ is* stable *for system* (A.1) *if for every $\varepsilon > 0$, there exists a $\delta > 0$ such that $|x_0|_{\mathcal{A}} \leq \delta$ implies that $|x(t)|_{\mathcal{A}} \leq \varepsilon$ for all $t \in \mathbb{I}_{\geq 0}$. It is* globally asymptotically stable *if it is stable and $\lim_{t \to \infty} |x(t)|_{\mathcal{A}} = 0$ for all $x_0 \in \mathbb{R}^n$.*

The above definition can also be adapted to the case where the system state x is constrained to lie in some positively invariant set $\mathbb{X} \subseteq \mathbb{R}^n$. Namely, a closed and positively invariant set $\mathcal{A} \subseteq \mathbb{X}$ is *asymptotically stable with region of attraction $\mathbb{X}$* if (i) $\lim_{t \to \infty} |x(t)|_{\mathcal{A}} = 0$ for all $x_0 \in \mathbb{X}$ and (ii) for every $\varepsilon > 0$, there exists a $\delta > 0$ such that $|x_0|_{\mathcal{A}} \leq \delta$ and $x_0 \in \mathbb{X}$ imply that $|x(t)|_{\mathcal{A}} \leq \varepsilon$ for all $t \in \mathbb{I}_{\geq 0}$. Furthermore, note that a special case of Definition A.1 is asymptotic stability of an equilibrium point x^* of system (A.1), i.e., $\mathcal{A} = \{x^*\}$.

As is shown in (Jiang and Wang, 2002, Proposition 3.2), in case that f is continuous and $\mathcal{A}$ is compact, global asymptotic stability as given in Definition A.1 is equivalent to the following definition based on[1] $\mathcal{KL}$ functions.

Definition A.2. *A closed and positively invariant set $\mathcal{A} \subseteq \mathbb{R}^n$ is* globally asymptotically stable *for system* (A.1) *if there exists a function $\beta \in \mathcal{KL}$ such that $|x(t)|_{\mathcal{A}} \leq \beta(|x_0|_{\mathcal{A}}, t)$ for all $t \in \mathbb{I}_{\geq 0}$ and all $x_0 \in \mathbb{R}^n$.*

On the other hand, if f is not continuous or $\mathcal{A}$ is not compact, Definitions A.1 and A.2 are not equivalent, but the latter is stronger; examples for this fact can be found in Jiang and Wang (2002) and Rawlings and Mayne (2012). In model predictive control, the closed-loop system (2.6) is not necessarily continuous, as the control input can be discontinuous (see, e.g., Example 2.8 in Rawlings and Mayne (2009)). Nevertheless, one can show that

---

[1] A function $\beta : \mathbb{R}_{\geq 0} \times \mathbb{R}_{\geq 0} \to \mathbb{R}_{\geq 0}$ is a class $\mathcal{KL}$ function (or $\beta \in \mathcal{KL}$ for short), if $\beta(\cdot, t) \in \mathcal{K}$ for each fixed $t \in \mathbb{R}_{\geq 0}$, and $\beta(r, \cdot)$ is decreasing with $\lim_{t \to \infty} \beta(r, t) = 0$ for each fixed $r \in \mathbb{R}_{\geq 0}$.

standard MPC stability results are still valid under the stronger definition of asymptotic stability, i.e., Definition A.2, in case that the set $\mathcal{A}$ is compact (Rawlings and Mayne, 2012); this is also true for the presented stability results in Chapter 3. On the other hand, if the set $\mathcal{A}$ is unbounded (as can, e.g., be the case in consensus and synchronization problems), the stability results established in Chapter 3 are only valid for the definition of asymptotic stability according to Definition A.1. Finally, we note that if Assumption 3.8 is strengthened to hold for all $x \in \mathbb{X}_N$ instead for all $x \in \mathbb{X}^f$, then the stability results of Chapter 3 are also valid for unbounded sets $\mathcal{A}$ with the stronger definition of asymptotic stability according to Definition A.2.

A.2 Asymptotic and essential convergence

One requirement in the definition of asymptotic stability (see Definition A.1) is that the state x of system (A.1) asymptotically converges to the set $\mathcal{A}$, i.e., $\lim_{t\to\infty} |x(t)|_{\mathcal{A}} = 0$ (or $\lim_{t\to\infty} |x(t) - x^*| = 0$ in the special case of an equilibrium point, i.e., $\mathcal{A} = \{x^*\}$). In some parts of this thesis, also a slightly weaker notion than asymptotic convergence of a sequence is used, namely *essential* convergence (see Section 4.2.2 and Chapter 5). In accordance with Angeli et al. (2011), we define this notion as follows.

Definition A.3. *A sequence* $\boldsymbol{v} : \mathbb{I}_{\geq 0} \to \mathbb{R}^n$ *is essentially converging to* $\bar{v} \in \mathbb{R}^n$ *if the following is satisfied for all* $\varepsilon > 0$:

$$\limsup_{T\to+\infty} \frac{card(\{0 \leq t \leq T : |v(t) - \bar{v}| \geq \varepsilon\})}{T+1} = 0.$$

Note that essential convergence is slightly weaker than asymptotic convergence. Namely, while asymptotic convergence clearly implies essential convergence, the converse is not necessarily true. For example, the sequence $\boldsymbol{v} : \mathbb{I}_{\geq 0} \to \mathbb{R}$ with $v(t) = 1$ if $t = 2^n$ for $n \in \mathbb{I}_{\geq 0}$, and $v(t) = 0$ otherwise, essentially converges to 0 while it clearly does not asymptotically converge to 0, see (Angeli et al., 2011).

Appendix B

Nonlinear programming

In this chapter, we give a brief overview on relevant results from nonlinear programming which are needed in Section 4.1.2; this overview is partially taken from Section II in (Müller et al., 2014d). Namely, there we consider several different optimization problems $\mathcal{P}(y, f_0, h, g)$ of the form

$$\underset{y}{\text{minimize}} \ f_0(y) \tag{B.1}$$

$$\text{subject to } h(y) = 0, \ g(y) \leq 0,$$

where $y \in \mathbb{R}^{n_y}$ and the functions $f_0 : \mathbb{R}^{n_y} \to \mathbb{R}$, $h : \mathbb{R}^{n_y} \to \mathbb{R}^{n_h}$ and $g : \mathbb{R}^{n_y} \to \mathbb{R}^{n_g}$ are assumed to be twice continuously differentiable. For every feasible point y, denote by $A(y) := \{1 \leq j \leq n_g : g_j(y) = 0\}$ the set of active inequality constraints at y. We say that a feasible point y is *regular* (Bertsekas, 1995), if $\nabla_y h_i(y)$, $1 \leq i \leq n_h$, and $\nabla_y g_j(y)$, $j \in A(y)$, are linearly independent. We then have the following first-order necessary conditions for optimality, known as the Karush-Kuhn-Tucker (KKT) conditions:

Proposition B.1 (Bertsekas (1995), Proposition 3.3.1)**.** *Suppose that y^* is regular and a local minimizer of problem $\mathcal{P}$. Then there exist unique Lagrange multipliers $\mu \in \mathbb{R}^{n_h}$ and $\nu \in \mathbb{R}^{n_g}$ such that the following holds:*

$$\nabla_y f_0(y^*) + \sum_{i=1}^{n_h} \mu_i \nabla_y h_i(y^*) + \sum_{j=1}^{n_g} \nu_j \nabla_y g_j(y^*) = 0, \tag{B.2}$$

$$\nu_j \geq 0, \quad \nu_j g_j(y^*) = 0, \quad 1 \leq j \leq n_g. \tag{B.3}$$

Furthermore, a KKT point (y^*, μ, ν) is said to satisfy the *strong second order sufficiency condition* (Jittorntrum, 1984; Robinson, 1980) if

$$w^T \Big(\nabla_y^2 f_0(y^*) + \sum_{i=1}^{n_h} \mu_i \nabla_y^2 h_i(y^*) + \sum_{j=1}^{n_g} \nu_j \nabla_y^2 g_j(y^*) \Big) w > 0 \tag{B.4}$$

for all $w \neq 0$ such that $\nabla_y h_i(y^*)w = 0$ for all $1 \leq i \leq n_h$ and $\nabla_y g_j(y^*)w = 0$ for all $j \in A(y^*)$ such that $\nu_j > 0$.

Proposition B.2 (Bertsekas (1995); Jittorntrum (1984))**.** *Suppose that y^* is a feasible point of problem $\mathcal{P}$ which is regular and together with some (μ, ν) satisfies the KKT conditions (B.2)-(B.3) as well as the strong second order sufficiency condition (B.4). Then y^* is a strict local minimizer of problem $\mathcal{P}$.*

Remark B.3. *In order for Proposition B.2 to hold, it suffices that a slightly weaker condition than the strong second order sufficiency condition holds. Namely, (B.4) has to hold only for such w which in addition to the above requirements also fulfill $\nabla_y g_j(y^*)w \leq 0$, for all j such that $j \in A(y^*)$ and $\nu_j = 0$ (Bertsekas, 1995, p.291). In Section 4.1.2, we use the strong second order sufficiency condition as it allows us to apply certain sensitivity results also in the case where strict complementarity (i.e., $\nu_j > 0$ for all $j \in A(y^*)$) does not hold (Jittorntrum, 1984; Robinson, 1980).*

In case that problem $\mathcal{P}$ is convex, i.e., in (B.1) the functions f_0 and g are convex and h is affine in y, much stronger statements can be obtained. For problem $\mathcal{P}$, *Slater's condition* (Boyd and Vandenberghe, 2004) is satisfied if there exists a point y which is strictly feasible, i.e., such that $h(y) = 0$ and $g(y) < 0$. With this constraint qualification, the following result can be obtained.

Proposition B.4 (Boyd and Vandenberghe (2004)). *Suppose that problem $\mathcal{P}$ is convex. Then y^* is a global minimizer of problem $\mathcal{P}$ if together with some (μ, ν) it satisfies the KKT conditions (B.2)-(B.3). The KKT conditions are also necessary in case that Slater's condition is satisfied for problem $\mathcal{P}$. Furthermore, if f_0 is strictly convex, the global minimizer y^* is unique.*

Bibliography

R. Amrit, J. B. Rawlings, and D. Angeli. Economic optimization using model predictive control with a terminal cost. *Annual Reviews in Control*, 35(2):178–186, 2011.

D. Angeli and J. B. Rawlings. Receding horizon cost optimization and control for nonlinear plants. In *Proceedings of the 8th IFAC Symposium on Nonlinear Control Systems*, pages 1217–1223, 2010.

D. Angeli, R. Amrit, and J. B. Rawlings. Receding horizon cost optimization for overly constrained nonlinear plants. In *Proceedings of the 48th IEEE Conference on Decision and Control and 28th Chinese Control Conference*, pages 7972–7977, 2009.

D. Angeli, R. Amrit, and J. B. Rawlings. Enforcing convergence in nonlinear economic MPC. In *Proceedings of the 50th IEEE Conference on Decision and Control and European Control Conference*, pages 3387–3391, 2011.

D. Angeli, R. Amrit, and J. B. Rawlings. On average performance and stability of economic model predictive control. *IEEE Transactions on Automatic Control*, 57(7):1615–1626, 2012.

J. P. Aubin and A. Cellina. *Differential Inclusions. Set-Valued Maps and Viability Theory*. Springer, 1984.

T. Backx, O. Bosgra, and W. Marquardt. Integration of model predictive control and optimization of processes. In *Proceedings of the IFAC Symposium on Advanced Control of Chemical Processes*, pages 249–260, 2000.

J. E. Bailey, F. J. M. Horn, and R. C. Lin. Cyclic operation of reaction systems: Effects of heat and mass transfer resistance. *AIChE Journal*, 17(4):818–825, 1971.

F. A. Bayer, M. A. Müller, and F. Allgöwer. Tube-based robust economic model predictive control. *Journal of Process Control*, 24(8):1237–1246, 2014.

D. P. Bertsekas. *Nonlinear Programming*. Athena Scientific, Belmont, Massachusetts, 1995.

S. Bittanti and G. Guardabassi. Optimal periodic control and periodic systems analysis: an overview. In *Proceedings of the 25th IEEE Conference on Decision and Control*, pages 1417–1423, 1986.

S. Bittanti, G. Fronza, and G. Guardabassi. Periodic control: A frequency domain approach. *IEEE Transactions on Automatic Control*, 18(1):33–38, 1973.

C. Böhm. *Predictive Control using Semi-definite Programming - Efficient Approaches for Periodic Systems and Lur'e Systems*. PhD thesis, University of Stuttgart, 2011.

S. Boyd and L. Vandenberghe. *Convex Optimization*. Cambridge University Press, New York, NY, USA, 2004.

S. Boyd, L. El Ghaoui, E. Feron, and V. Balakrishnan. *Linear Matrix Inequalities in System and Control Theory*, volume 15 of *Studies in Applied Mathematics*. SIAM, Philadelphia, PA, 1994.

M. S. Branicky. Multiple Lyapunov functions and other analysis tools for switched and hybrid systems. *IEEE Transactions on Automatic Control*, 43(4):475–482, 1998.

M. Bürger, G. Notarstefano, and F. Allgöwer. A polyhedral approximation framework for convex and robust distributed optimization. *IEEE Transactions on Automatic Control*, 59(2):384–395, 2014.

C. I. Byrnes and W. Lin. Losslessness, feedback equivalence, and the global stabilization of discrete-time nonlinear systems. *IEEE Transactions on Automatic Control*, 39(1):83–98, 1994.

E. F. Camacho and C. Bordons. *Model Predictive Control*. Springer, London, 2004.

H. Chen and F. Allgöwer. A quasi-infinite horizon nonlinear model predictive control scheme with guaranteed stability. *Automatica*, 34(10):1205–1217, 1998.

X. Chen, M. Heidarinejad, J. Liu, and P. D. Christofides. Distributed economic MPC: Application to a nonlinear chemical process network. *Journal of Process Control*, 22(4):689–699, 2012.

P. D. Christofides, R. Scattolini, D. Muñoz de la Peña, and J. Liu. Distributed model predictive control: A tutorial review and future research directions. *Computers & Chemical Engineering*, 51:21–41, 2013.

B. Chu, S. Duncan, A. Papachristodoulou, and C. Hepburn. Using economic model predictive control to design sustainable policies for mitigating climate change. In *Proceedings of the 51st IEEE Conference on Decision and Control*, pages 406–411, 2012.

C. Conte, N. R. Voellmy, M. N. Zeilinger, M. Morari, and C. N. Jones. Distributed synthesis and control of constrained linear systems. In *Proceedings of the American Control Conference*, pages 6017–6022, 2012.

T. Damm, L. Grüne, M. Stieler, and K. Worthmann. An exponential turnpike theorem for dissipative discrete time optimal control problems. *SIAM Journal on Control and Optimization*, 52(3):1935–1957, 2014.

M. L. Darby and M. Nikolaou. MPC: Current practice and challenges. *Control Engineering Practice*, 20:328–342, 2012.

M. Diehl, R. Amrit, and J. B. Rawlings. A Lyapunov function for economic optimizing model predictive control. *IEEE Transactions on Automatic Control*, 56(3):703–707, 2011.

M. D. Doan, T. Keviczky, and B. De Schutter. A distributed optimization-based approach for hierarchical MPC of large-scale systems with coupled dynamics and constraints. In *Proceedings of the 50th IEEE Conference on Decision and Control and European Control Conference*, pages 5236–5241, 2011.

P. A. A. Driessen, R. M. Hermans, and P. P. J. van den Bosch. Distributed economic model predictive control of networks in competitive environments. In *Proceedings of the 51st IEEE Conference on Decision and Control*, pages 266–271, 2012.

W. B. Dunbar and D. S. Caveney. Distributed receding horizon control of vehicle platoons: Stability and string stability. *IEEE Transactions on Automatic Control*, 57(3):620–633, 2012.

W. B. Dunbar and R. M. Murray. Distributed receding horizon control for multi-vehicle formation stabilization. *Automatica*, 42(4):549–558, 2006.

M. Ellis and P. D. Christofides. Economic model predictive control with time-varying objective function for nonlinear process systems. *AIChE Journal*, 60(2):507–519, 2014.

S. Engell. Feedback control for optimal process operation. *Journal of Process Control*, 17 (3):203 – 219, 2007.

L. Fagiano and A. R. Teel. Model predictive control with generalized terminal constraint. In *Proceedings of the 4th IFAC Nonlinear Model Predictive Control Conference*, pages 299–304, 2012.

L. Fagiano and A. R. Teel. Generalized terminal state constraint for model predictive control. *Automatica*, 49(9):2622–2631, 2013.

M. Farina and R. Scattolini. Distributed predictive control: A non-cooperative algorithm with neighbor-to-neighbor communication for linear systems. *Automatica*, 48(6):1088–1096, 2012.

T. Faulwasser. *Optimization-based solutions to constrained trajectory-tracking and path-following problems*. Contributions in Systems Theory and Automatic Control, no. 3. Shaker, Aachen, Germany, 2013. doi: 10.2370/9783844015942.

T. Faulwasser and R. Findeisen. Nonlinear model predictive path-following control. In L. Magni, D. M. Raimondo, and F. Allgöwer, editors, *Nonlinear Model Predictive Control - Towards New Challenging Applications*, Lecture Notes in Control and Information Sciences, pages 335–343. Springer, 2009.

A. Ferramosca. *Model predictive control of systems with changing setpoints*. PhD thesis, Universidad de Sevilla, 2011.

A. Ferramosca, D. Limon, I. Alvarado, T. Alamo, and E. F. Camacho. MPC for tracking with optimal closed-loop performance. *Automatica*, 45(8):1975–1978, 2009.

A. Ferramosca, J. B. Rawlings, D. Limon, and E. F. Camacho. Economic MPC for a changing economic criterion. In *Proceedings of the 49th IEEE Conference on Decision and Control*, pages 6131–6136, 2010.

G. Ferrari-Trecate, L. Galbusera, M. P. E. Marciandi, and R. Scattolini. Model predictive control schemes for consensus in multi-agent systems with single- and double-integrator dynamics. *IEEE Transactions on Automatic Control*, 54(11):2560–2572, 2009.

A. V. Fiacco and Y. Ishizuka. Sensitivity and stability analysis for nonlinear programming. *Annals of Operations Research*, 27(1):215–235, 1990.

R. Findeisen, L. Imsland, F. Allgöwer, and B. A. Foss. State and output feedback nonlinear model predictive control: An overview. *European Journal of Control*, 9(2-3):190–206, 2003.

R. Findeisen, L. Grüne, J. Pannek, and P. Varutti. Robustness of prediction based delay compensation for nonlinear systems. In *Proceedings of the 18th IFAC World Congress*, pages 203–208, 2011.

P. Finsler. Über das Vorkommen definiter und semidefiniter Formen in Scharen quadratischer Formen. *Comentarii Mathematici Helvetici*, 9(1):188–192, 1937.

F. A. C. C. Fontes. A general framework to design stabilizing nonlinear model predictive controllers. *Systems & Control Letters*, 42(2):127–143, 2001.

E. Franco, L. Magni, T. Parisini, M. M. Polycarpou, and D. M. Raimondo. Cooperative constrained control of distributed agents with nonlinear dynamics and delayed information exchange: A stabilizing receding-horizon approach. *IEEE Transactions on Automatic Control*, 53(1):324–338, 2008.

A. Gautam, Y.-C. Chu, and Y. C. Soh. Robust H_∞ receding horizon control for a class of coordinated control problems involving dynamically decoupled subsystems. *IEEE Transactions on Automatic Control*, 59(1):134–149, Jan 2014.

E. G. Gilbert. Optimal periodic control: A general theory of necessary conditions. *SIAM Jornal on Control and Optimization*, 15(5):717–746, 1977.

P. Giselsson and A. Rantzer. On feasibility, stability and performance in distributed model predictive control. *IEEE Transactions on Automatic Control*, 59(4):1031–1036, 2014.

C. Godsil and G. Royle. *Algebraic Graph Theory*. Springer, 2001.

R. Goebel, R. G. Sanfelice, and A. R. Teel. *Hybrid Dynamical Systems: Modeling, Stability, and Robustness*. Princeton University Press, 2012.

R. Gondhalekar and C. N. Jones. MPC of constrained discrete-time linear periodic systems — a framework for asynchronous control: Strong feasibility, stability and optimality via periodic invariance. *Automatica*, 47(2):326 – 333, 2011.

S. Gros. An economic NMPC formulation for wind turbine control. In *Proceedings of the 52nd IEEE Conference on Decision and Control*, pages 1001–1006, 2013.

D. Groß and O. Stursberg. On the convergence rate of a Jacobi algorithm for cooperative distributed MPC. In *Proceedings of the 52nd IEEE Conference on Decision and Control*, pages 1508–1513, 2013.

L. Grüne. Analysis and design of unconstrained nonlinear MPC schemes for finite and infinite dimensional systems. *SIAM Journal on Control and Optimization*, 48(2):1206–1228, 2009.

L. Grüne. Economic receding horizon control without terminal constraints. *Automatica*, 49(3):725–734, 2013.

L. Grüne and J. Pannek. *Nonlinear Model Predictive Control. Theory and Algorithms*. Springer, London, 2011.

L. Grüne and M. Stieler. Asymptotic stability and transient optimality of economic MPC without terminal constraints. *Journal of Process Control*, 24(8):1187–1196, 2014.

L. Grüne and K. Worthmann. A distributed NMPC scheme without stabilizing terminal constraints. In R. Johansson and A. Rantzer, editors, *Distributed Decision Making and Control*, volume 417 of *Lecture Notes in Control and Information Sciences*, pages 261–287. Springer, 2012.

L. Grüne and M. Zanon. Periodic optimal control, dissipativity and MPC. In *Proceedings of the 21st International Symposium on Mathematical Theory of Networks and Systems*, pages 1804–1807, 2014.

L. Grüne, F. Allgöwer, R. Findeisen, J. Fischer, D. Groß, U. D. Hanebeck, B. Kern, M. A. Müller, J. Pannek, M. Reble, O. Stursberg, P. Varutti, and K. Worthmann. Distributed and networked model-predictive control. In *J. Lunze, editor, Control Theory of Digitally Networked Systems*, pages 111–167. Springer, 2014.

M. M. Halldórsson. A still better performance guarantee for approximate graph coloring. *Information Processing Letters*, 45(1):19–23, 1993.

M. Heidarinejad. *Economic and Distributed Model Predictive Control of Nonlinear Systems*. PhD thesis, University of California, Los Angeles, 2012.

M. Heidarinejad, J. Liu, and P. D. Christofides. Economic model predictive control of nonlinear process systems using Lyapunov techniques. *AIChE Journal*, 58(3):855–870, 2012.

R. A. Horn and C. R. Johnson. *Topics in matrix analysis*. Cambridge University Press, 1991.

B. Houska, H. J. Ferreau, and M. Diehl. Acado toolkit – an open-source framework for automatic control and dynamic optimization. *Optimal Control Applications and Methods*, 32(3):298–312, 2011.

T. G. Hovgaard, L. F. S. Larsen, and J. B. Jørgensen. Robust economic MPC for a power management scenario with uncertainties. In *Proceedings of the 50th IEEE Conference on Decision and Control and European Control Conference*, pages 1515–1520, 2011.

T. G. Hovgaard, L. F. S. Larsen, K. Edlund, and J. B. Jørgensen. Model predictive control technologies for efficient and flexible power consumption in refrigeration systems. *Energy*, 44(1):105 – 116, 2012.

R. Huang, E. Harinath, and L. T. Biegler. Lyapunov stability of economically oriented NMPC for cyclic processes. *Journal of Process Control*, 21(4):501–509, 2011.

R. Huang, L. T. Biegler, and E. Harinath. Robust stability of economically oriented infinite horizon NMPC that include cyclic processes. *Journal of Process Control*, 22(1):51–59, 2012.

A. Jadbabaie, J. Lin, and A. S. Morse. Coordination of groups of mobile autonomous agents using nearest neighbor rules. *IEEE Transactions on Automatic Control*, 48(6): 988–1001, 2003.

Z.-P. Jiang and Y. Wang. A converse Lyapunov theorem for discrete-time systems with disturbances. *Systems & Control Letters*, 45(1):49–58, 2002.

K. Jittorntrum. Solution point differentiability without strict complementarity in nonlinear programming. In A. V. Fiacco, editor, *Sensitivity, Stability and Parametric Analysis*, volume 21 of *Mathematical Programming Studies*, pages 127–138. Springer Berlin Heidelberg, 1984.

B. J. Johansson, A. Speranzon, M. Johansson, and K. H. Johansson. Distributed model predictive consensus. In *Proceedings of the 17th International Symposium on Mathematical Theory of Networks and Systems*, pages 2438–2444, 2006.

J. V. Kadam and W. Marquardt. Integration of economical optimization and control for intentionally transient process operation. In R. Findeisen, F. Allgöwer, and L. T. Biegler, editors, *Assessment and Future Directions of Nonlinear Model Predictive Control*, Lecture Notes in Control and Information Sciences, pages 419–434. Springer, 2007.

T. Keviczky and K. H. Johansson. A study on distributed model predictive consensus. In *Proceedings of the 17th IFAC World Congress*, pages 1516–1521, 2008.

T. Keviczky, F. Borrelli, and G. J. Balas. Decentralized receding horizon control for large scale dynamically decoupled systems. *Automatica*, 42(12):2105–2115, 2006.

H. K. Khalil. *Nonlinear Systems*. Prentice-Hall, Upper Saddle River, NJ, 2002.

M. Kögel and R. Findeisen. Cooperative distributed MPC using the alternating direction multiplier method. In *Proceedings of the 8th IFAC Symposium on Advanced Control of Chemical Processes*, pages 445–450, 2012.

M. V. Kothare, V. Balakrishnan, and M. Morari. Robust constrained model predictive control using linear matrix inequalities. *Automatica*, 32(10):1361–1379, 1996.

M. Lazar, W. P. M. H. Heemels, S. Weiland, and A. Bemporad. Stabilizing model predictive control of hybrid systems. *IEEE Transactions on Automatic Control*, 51(11):1813–1818, 2006.

J. Lee and D. Angeli. Cooperative distributed model predictive control for linear plants subject to convex economic objectives. In *Proceedings of the 50th IEEE Conference on Decision and Control and European Control Conference*, pages 3434–3439, 2011.

J. Lee and D. Angeli. Distributed cooperative nonlinear economic MPC. In *Proceedings of the 20th International Symposium on Mathematical Theory of Networks and Systems*, 2012.

J. Lee, J.-S. Kim, H. Song, and H. Shim. A constrained consensus problem using MPC. *International Journal of Control, Automation and Systems*, 9(5):952–957, 2011.

D. Liberzon. *Switching in Systems and Control*. Birkhäuser, Boston, MA, 2003.

D. Limon, I. Alvarado, T. Alamo, and E. F. Camacho. MPC for tracking piecewise constant references for constrained linear systems. *Automatica*, 44(9):2382–2387, 2008.

D. Limon, T. Alamo, D. M. Raimondo, D. Muñoz de la Peña, J. M. Bravo, A. Ferramosca, and E. F. Camacho. Input-to-state stability: A unifying framework for robust model predictive control. In L. Magni, D. M. Raimondo, and F. Allgöwer, editors, *Nonlinear Model Predictive Control - Towards New Challenging Applications*, Lecture Notes in Control and Information Sciences, pages 1–26. Springer, 2009.

J. Liu, D. Muñoz de la Peña, and P. D. Christofides. Distributed model predictive control of nonlinear systems subject to asynchronous and delayed measurements. *Automatica*, 46(1):52–61, 2010.

J. M. Maestre and R. R. Negenborn, editors. *Distributed Model Predictive Control Made Easy*. Springer, 2014.

J. M. Maestre, D. Muñoz de la Peña, E. F. Camacho, and T. Alamo. Distributed model predictive control based on agent negotiation. *Journal of Process Control*, 21(5):685–697, 2011.

J. P. Maree and L. Imsland. Multi-objective predictive control for non steady-state operation. In *Proceedings of the European Control Conference*, pages 1541–1546, July 2013.

D. Q. Mayne, J. B. Rawlings, C. V. Rao, and P. O. M. Scokaert. Constrained model predictive control: Stability and optimality. *Automatica*, 36(6):789–814, 2000.

D. Q. Mayne, M. M. Seron, and S. V. Raković. Robust model predictive control of constrained linear systems with bounded disturbances. *Automatica*, 41(2):219–224, 2005.

P. Mhaskar, N. H. El-Farra, and P. D. Christofides. Predictive control of switched nonlinear systems with scheduled mode transitions. *IEEE Transactions on Automatic Control*, 50 (11):1670 – 1680, 2005.

P. Mhaskar, N. H. El-Farra, and P. D. Christofides. Stabilization of nonlinear systems with state and control constraints using Lyapunov-based predictive control. *Systems & Control Letters*, 55(8):650 – 659, 2006.

M. A. Müller and F. Allgöwer. Robustness of steady-state optimality in economic model predictive control. In *Proceedings of the 51st IEEE Conference on Decision and Control*, pages 1011–1016, 2012.

M. A. Müller and F. Allgöwer. Distributed MPC for consensus and synchronization. In *J. M. Maestre, R. Negenborn, editors, Distributed Model Predictive Control Made Easy*, pages 89–100. Springer, 2014a.

M. A. Müller and F. Allgöwer. Distributed economic MPC: a framework for cooperative control problems. In *Proceedings of the 19th IFAC World Congress*, pages 1029–1034, 2014b.

M. A. Müller, M. Reble, and F. Allgöwer. A general distributed MPC framework for cooperative control. In *Proceedings of the 18th IFAC World Congress*, pages 7987–7992, 2011.

M. A. Müller, P. Martius, and F. Allgöwer. Model predictive control of switched nonlinear systems under average dwell-time. *Journal of Process Control*, 22(9):1702–1710, 2012a.

M. A. Müller, M. Reble, and F. Allgöwer. Cooperative control of dynamically decoupled systems via distributed model predictive control. *International Journal of Robust and Nonlinear Control*, 22(12):1376–1397, 2012b.

M. A. Müller, B. Schürmann, and F. Allgöwer. Robust cooperative control of dynamically decoupled systems via distributed MPC. In *Proceedings of the 4th IFAC Nonlinear Model Predictive Control Conference*, pages 412–417, 2012c.

M. A. Müller, D. Angeli, and F. Allgöwer. On convergence of averagely constrained economic MPC and necessity of dissipativity for optimal steady-state operation. In *Proceedings of the American Control Conference*, pages 3147–3152, 2013a.

M. A. Müller, D. Angeli, and F. Allgöwer. Economic model predictive control with transient average constraints. In *Proceedings of the 52nd IEEE Conference on Decision and Control*, pages 5119–5124, 2013b.

M. A. Müller, D. Angeli, and F. Allgöwer. Economic model predictive control with self-tuning terminal weight. In *Proceedings of the European Control Conference*, pages 2044–2049, 2013c.

M. A. Müller, D. Angeli, and F. Allgöwer. Economic model predictive control with self-tuning terminal cost. *European Journal of Control*, 19(5):408–416, 2013d.

M. A. Müller, D. Angeli, and F. Allgöwer. Performance analysis of economic MPC with self-tuning terminal cost. In *Proceedings of the American Control Conference*, pages 2845–2850, 2014a.

M. A. Müller, D. Angeli, and F. Allgöwer. Transient average constraints in economic model predictive control. *Automatica*, 2014b. To appear.

M. A. Müller, D. Angeli, and F. Allgöwer. On necessity and robustness of dissipativity in economic model predictive control. *IEEE Transactions on Automatic Control*, 2014c. To appear.

M. A. Müller, D. Angeli, and F. Allgöwer. Additional material to the paper 'On necessity and robustness of dissipativity in economic model predictive control'. 2014d. Available at: http://www.simtech.uni-stuttgart.de/publikationen/prints.php?ID=903.

M. A. Müller, D. Angeli, and F. Allgöwer. On the performance of economic model predictive control with self-tuning terminal cost. *Journal of Process Control*, 24(8):1179–1186, 2014e.

M. A. Müller, D. Angeli, F. Allgöwer, R. Amrit, and J. B. Rawlings. Convergence in economic model predictive control with average constraints. *Automatica*, 2014f. To appear.

R. Olfati-Saber and R. M. Murray. Consensus problems in networks of agents with switching topology and time-delays. *IEEE Transactions on Automatic Control*, 49(9):1520–1533, 2004.

R. Olfati-Saber, J. A. Fax, and R. M. Murray. Consensus and cooperation in networked multi-agent systems. *Proceedings of the IEEE*, 95(1):215–233, 2007.

B. P. Omell and D. J. Chmielewski. Application of infinite-horizon EMPC to IGCC dispatch. In *Proceedings of the American Control Conference*, pages 5358–5363, 2013.

G. Pannocchia, J. B. Rawlings, and S. J. Wright. Conditions under which suboptimal nonlinear MPC is inherently robust. *Systems & Control Letters*, 60(9):747 – 755, 2011.

G. Pin and T. Parisini. Networked predictive control of uncertain constrained nonlinear systems: Recursive feasibility and input-to-state stability analysis. *IEEE Transactions on Automatic Control*, 56(1):72–87, 2011.

S. J. Qin and T. A. Badgwell. A survey of industrial model predictive control technology. *Control Engineering Practice*, 11:733–764, 2003.

D. M. Raimondo, D. Limon, M. Lazar, L. Magni, and E. F. Camacho. Min-max model predictive control of nonlinear systems: a unifying overview on stability. *European Journal of Control*, 15(1):5–21, 2009.

S. V. Raković, B. Kouvaritakis, M. Cannon, C. Panos, and R. Findeisen. Parameterized tube model predictive control. *IEEE Transactions on Automatic Control*, 57(11):2746–2761, 2012.

J. B. Rawlings and R. Amrit. Optimizing process economic performance using model predictive control. In L. Magni, D. M. Raimondo, and F. Allgöwer, editors, *Nonlinear Model Predictive Control - Towards New Challenging Applications*, Lecture Notes in Control and Information Sciences, pages 119–138. Springer, 2009.

J. B. Rawlings and D. Q. Mayne. *Model Predictive Control: Theory and Design*. Nob Hill Publishing, Madison, WI, 2009.

J. B. Rawlings and D. Q. Mayne. Postface to "Model Predictive Control: Theory and Design". 2012. Available at: http://jbrwww.che.wisc.edu/home/jbraw/mpc/postface.pdf.

J. B. Rawlings, D. Angli, and C. N. Bates. Fundamentals of economic model predictive control. In *Proceedings of the 51st IEEE Conference on Decision and Control*, pages 3851–3861, 2012.

M. Reble. *Model Predictive Control for Nonlinear Continuous-Time Systems with and without Time-Delays*. PhD thesis, University of Stuttgart, 2013.

M. Reble and F. Allgöwer. Unconstrained model predictive control and suboptimality estimates for nonlinear continuous-time systems. *Automatica*, 48(8):1812–1817, 2012.

A. Richards and J. P. How. Robust distributed model predictive control. *International Journal of Control*, 80(9):1517–1531, 2007.

S. Riverso, M. Farina, and G. Ferrari-Trecate. Plug-and-play decentralized model predictive control for linear systems. *IEEE Transactions on Automatic Control*, 58(10):2608–2614, 2013.

S. M. Robinson. Strongly regular generalized equations. *Mathematics of Operations Research*, 5(1):43–62, 1980.

C. Rueger. *Soli Deo Gloria. Johann Sebastian Bach*. Erika Klopp Verlag, Berlin, 1985.

A. Ruszczyński. *Nonlinear Optimization*. Princeton University Press, 2006.

R. Scattolini. Architectures for distributed and hierarchical model predictive control - a review. *Journal of Process Control*, 19(5):723–731, 2009.

R. M. Schaich. Distributed model predictive control for networks with time-varying communication topology. Master thesis, University of Stuttgart, 2013.

R. M. Schaich, M. A. Müller, and F. Allgöwer. A distributed model predictive control scheme for networks with communication failure. In *Proceedings of the 19th IFAC World Congress*, pages 12004–12009, 2014.

H. Scheu and W. Marquardt. Sensitivity-based coordination in distributed model predictive control. *Journal of Process Control*, 21(5):715–728, 2011.

B. Schürmann. Robustifying a distributed MPC approach for cooperative control of dynamically decoupled systems. Bachelor thesis, University of Stuttgart, 2012.

S. E. Shafiei, J. Stoustrup, and H. Rasmussen. A supervisory control approach in economic MPC design for refrigeration systems. In *Proceedings of the European Control Conference*, pages 1565–1570, 2013.

E. D. Sontag. *Mathematical Control Theory - Deterministic Finite Dimensional Systems*. Springer, New York, 2nd edition, 1998.

V. Spudić and M. Baotić. Fast coordinated model predictive control of large-scale distributed systems with single coupling constraint. In *Proceedings of the European Control Conference*, pages 2783–2788, 2013.

B. T. Stewart, A. N. Venkat, J. B. Rawlings, S. J. Wright, and G. Pannocchia. Cooperative distributed model predictive control. *Systems & Control Letters*, 59(8):460–469, 2010.

J. Stoustrup. Plug & play control: Control technology towards new challenges. *European Journal of Control*, 15(3-4):311–330, 2009.

P. Trodden and A. Richards. Distributed model predictive control of linear systems with persistent disturbances. *International Journal of Control*, 83(8):1653–1663, 2010.

P. Trodden and A. Richards. Cooperative distributed MPC of linear systems with coupled constraints. *Automatica*, 49(2):479 – 487, 2013.

P. Varutti, B. Kern, and R. Findeisen. Dissipativity-based distributed nonlinear predictive control for cascaded systems. In *Proceedings of the 8th IFAC Symposium on Advanced Control of Chemical Processes*, pages 439–444, 2012.

A. N. Venkat, J. B. Rawlings, and S. J. Wright. Stability and optimality of distributed model predictive control. In *Proceedings of the 44th IEEE Conference on Decision and Control and European Control Conference*, pages 6680–6685, 2005.

C. Wang and C.-J. Ong. Distributed model predictive control of dynamically decoupled systems with coupled cost. *Automatica*, 46(12):2053–2058, 2010.

G. Wang. Distributed economic model predictive control. Diploma thesis, University of Stuttgart, 2011.

P. Wieland. *From Static to Dynamic Couplings in Consensus and Synchronization among Identical and Non-Identical Systems*. PhD thesis, University of Stuttgart, 2010.

J. C. Willems. Dissipative dynamical systems - part i: General theory. *Archive for Rational Mechanics and Analysis*, 45(5):321–351, 1972.

I. J. Wolf, H. Scheu, and W. Marquardt. A hierarchical distributed economic NMPC architecture based on neighboring-extremal updates. In *Proceedings of the American Control Conference*, pages 4155–4160, 2012.

A. J. Wood and B. F. Wollenberg. *Power generation, operation, and control. Second edition.* John Wiley & Sons, 1996.

S. Yu. *Robust Model Predictive Control of Constrained Systems*. PhD thesis, University of Stuttgart, 2011.

M. Zanon, S. Gros, and M. Diehl. A Lyapunov function for periodic economic optimizing model predictive control. In *Proceedings of the 52nd IEEE Conference on Decision and Control*, pages 5107–5112, 2013.

V. M. Zavala and A. Flores-Tlacuahuac. Stability of multiobjective predictive control: A utopia-tracking approach. *Automatica*, 48(10):2627 – 2632, 2012.

D. Zelazo and M. Mesbahi. Edge agreement: Graph-theoretic performance bounds and passivity analysis. *IEEE Transactions on Automatic Control*, 56(3):544–555, 2011.

D. Zelazo, M. Bürger, and F. Allgöwer. A finite-time dual method for negotiation between dynamical systems. *SIAM Journal on Control and Optimization*, 51(1):172–194, 2013.

J. Zhan and X. Li. Consensus of sampled-data multi-agent networking systems via model predictive control. *Automatica*, 49(8):2502 – 2507, 2013.